ATMOSPHERIC TIDES

SYDNEY CHAPMAN

National Center for Atmospheric Research, Boulder, Colo.,
and Geophysical Institute, University of Alaska, College, Alaska, U.S.A.

RICHARD S. LINDZEN

Dept. of Geophysical Sciences, University of Chicago, Chicago, Ill. U.S.A.

ATMOSPHERIC TIDES

Thermal and Gravitational

D. REIDEL PUBLISHING COMPANY

DORDRECHT-HOLLAND

ISBN-13: 978-94-010-3401-2   e-ISBN-13: 978-94-010-3399-2
DOI: 10.1007/ 978-94-010-3399-2

© 1970. D. Reidel Publishing Company, Dordrecht, Holland
Softcover reprint of the hardcover 1st edition 1970

PREFACE

Everyone is familiar with the daily changes of air temperature. The barometer shows that these are accompanied by daily changes of mass distribution of the atmosphere, and consequently with daily motions of the air. In the tropics the daily pressure change is evident on the barographs; in temperate and higher latitudes it is not noticeable, being overwhelmed by cyclonic and anticyclonic pressure variations. There too, however, the daily change can be found by averaging the variations over many days; and the same process suffices to show that there is a still smaller *lunar* tide in the atmosphere, first sought by Laplace. Throughout nearly two centuries these 'tides', thermal and gravitational, have been extensively discussed in the periodical literature of science, although they are very minor phenomena at ground level. This monograph summarizes our present knowledge and theoretical understanding of them.

It is more than twenty years since the appearance of the one previous monograph on them – by Wilkes – and nearly a decade since they were last comprehensively reviewed, by Siebert. The intervening years have seen many additions to our knowledge of the state of the upper atmosphere, and of the tides there, on the basis of measurements by radio, rockets and satellites. Combined with progress in the dynamical theory, this has led to the definite abandonment of Kelvin's proposed resonance explanation of why the half-daily change of the barometer is so much more prominent than the 24-hourly component variation, although the latter is dominant in the air temperature changes that cause both these pressure variations. The main cause is the absorption of solar radiation by ozone in the stratosphere, aided by water vapor.

The dynamical theory indicates that the strength of the tides increases with height in the atmosphere, so that in the ionospheric regions they are no longer minor; there they are major aeronomic features, involving very strong winds, and tidal changes of height reckoned in kilometers. This has been well confirmed by observation. As now developed, the theory solves some (though not yet all) of the major traditional problems relating to the oscillations observed at ground level. The refinement of the theory involves a new generation of problems, ranging from the nature of the daily variations in the lowest few meters of the air near the ground, to the influence of factors of the environment in the atmosphere above 50 km – such as hydromagnetic forces, molecular viscosity and conduction, and the modified composition at high levels. The theory is even being extended to deal with thermal tides in the atmosphere of Mars. The shortcomings of the theory at present are indicated, and the likely directions of future progress are pointed out.

Chapter 1 gives a historical account of the subject, which has something of the flavor of a detective story in which the whodunit was the source of the semidiurnal pressure variation: when Kelvin proposed his hypothesis, the existence of the ozone layer, the chief cause, was unknown. Chapters 2S and 2L describe the main known facts about the various components of the solar thermal tide and the lunar gravitational tide, as revealed by the thermometer, barometer and anemometer, and as discussed and represented, graphically and mathematically, in various ways and by different authors. Chapter 3 summarizes the present state of the dynamical theory; it is by R. Lindzen, who also contributed the last three sections of Chapter 1; apart from this, Chapters 1 and 2 are by S. Chapman.

It is hoped that readers will benefit from the full subject index, author index, place index, and the Guide to and Classification of the Figures and Tables, on which much effort has been spent.

The typing assistance of Mrs. D. G. Fisher, Mrs. C. MacBride and Mrs. L. Reiffel is gratefully acknowledged.

SYDNEY CHAPMAN

RICHARD S. LINDZEN

July 15, 1969

TABLE OF CONTENTS

CHAPTER 1

INTRODUCTORY AND HISTORICAL

1.1. Introduction: Pytheas, Bacon, Newton and Laplace

The large-scale oscillations of the atmosphere, the subject of this review, are those produced (in two very different ways) by (a) the gravitational forces of the moon and sun, and (b) the thermal action of the sun. *Tide* here signifies oscillations excited gravitationally or thermally; *thermal* tide signifies the part excited thermally.

The sea tides, with rise and fall of the water twice daily on most open ocean coasts, have been known from time immemorial. But in the Mediterranean Sea, which may be regarded as a large lake, the tide is inconspicuous. The connection of the tides with the moon seems to have been first recognized and recorded, as far as available records go, by a famous explorer-mariner of Marseilles, Pytheas. Not long before 320 B.C. he made the first extensive voyage – of order 7000 miles or 11 000 km – westward out of the Mediterranean Sea and northward to Britain and beyond. He visited the Cornish tin mines and circumnavigated Britain. Thus he had occasion to observe the considerable tides on its coasts, and the daily regression of the times of high water, parallel to that of the time of the moon's transit. His writings have been lost, and are known only through quotations and allusions by later authors. Some of these were incredulous about his geographical discoveries, and scoffed at them; they found it hard to believe that lands in the latitudes of Britain could be habitable – knowing nothing of the Gulf Stream and its effect on the climate of western Europe. But for centuries geographers depended on Pytheas' data for information about northern countries (Hyde, 1947; Casson, 1959).

The occurrence, at many places, of high tide at about the time of the moon's passage across the meridian early prompted the idea that the moon exerts an *attraction* on the water.* But the occurrence of a second high tide when the moon is on or near the opposite meridian was a great puzzle to the few philosophers who thought about it. However, already about 1250 A.D. a rational (though wholly false) solution of this problem was attempted by the Franciscan friar Roger Bacon of Oxford. It was based on the Ptolemaic conception of the universe.

---

* Neckam, a learned English monk (1157–1217), who for a time was a professor at the University of Paris, wrote a book of general knowledge, *De Naturis Rerum*, in Latin, about 1190 A.D., which was published and edited, with copious comment, by Wright (1863). It is notable for containing the first known European reference to the mariner's compass. Mentioning the tides, he remarked that he was unable to resolve the vexed question as to their cause, but that the common belief was that they are due to the moon (Wright, 1863, p. xxvii).

This pictured the stars as lying on the inner surface of a crystal sphere (the *primum mobile*), centered on and daily revolving around the earth. The sun, moon and planets were thought to move in the space between this sphere and the earth. Bacon suggested that the moon emits rays of attraction; those that fall on the hemisphere facing the moon draw the water upward to create the tide there; those that miss the earth travel on to meet the crystal sphere, which reflects some of them to the opposite terrestrial hemisphere, raising the waters there and producing the opposite tide!

The true explanation of the tides was first indicated by Newton (1687a) in his *Principia Mathematica*. They are a consequence of the lunar and solar gravitational forces acting in accordance with his three laws of mechanics. His theory showed that the sun also produces a tide, which modulates the lunar tide, so that at full and new moon the tidal range is higher (the spring tides) and at the intervening epochs of half moon it is lower (the neap tides) than the average.

The solar sea tide can be explained in an approximate way as follows, treating the sun as an immovable mass S, with center O, round which the earth, of mass E and center C at distance r_S, revolves with angular velocity ω. The total centrifugal force $E\omega^2 r_S$ on the earth is balanced by the total gravitational attraction GSE/r_s^2, where G denotes the constant of gravitation. Hence $\omega^2 = GS/r_s^3$.

But over the earth's hemisphere nearer to the sun the gravitational attraction is greater than at C and the centrifugal acceleration is less, so that over this hemisphere there is a distribution of unbalanced upward force directed sunwards (at most points obliquely to the vertical), acting against the earth's own gravitational force. Over the opposite hemisphere the gravitational force of the sun is less and the centrifugal force is greater than at C, so that there is a distribution of unbalanced force obliquely upward there also. Particles free to move, like those of the sea and air, will do so under the action of these unbalanced forces; the water surface tends to become spheroidal, with the long axis along the line OC.

If the earth did not rotate, such a steady distribution of level could be attained; it is called the *solar equilibrium tide*, and is proportional to the gradient, at C, of the acceleration $\omega^2 r - GS/r^2$: that is, to $\omega^2 + 2GS/r_s^3$ or $3GS/r^3$.

Less simply, it can be shown that the moon's tidal force is similarly proportional to $3GM/r_M^3$, where M denotes the mass of the moon and r_M its distance from C. Thus the ratio of the tidal force of the moon to that of the sun is $(M/r_M^3)/(S/r_s^3)$. Because $M = \frac{4}{3}\pi\varrho_M a_M^3$ and $S = \frac{4}{3}\pi\varrho_S a_S^3$, where ϱ_M, ϱ_S denote the mean densities, and a_M, a_S the radii, of the moon and sun respectively, the ratio may be expressed as $\{(a_M/r_M)^3/(a_S/r_S)^3\}(\varrho_M/\varrho_S)$. Here the first factor is very nearly equal to 1, because (as seen at total solar eclipses) the sun and moon subtend almost equal angles at the earth. Hence the lunar/solar tidal ratio (2.15 for the principal terms) is very nearly equal to ϱ_M/ϱ_S, i.e. (3.34/1.41), or 2.37. Thus 'the moon rules the tides'. A more detailed discussion of tidal forcing is given in Section 3.4A.

The 'equilibrium tide', however, is only a theoretical conception (Lamb, 1932, p. 358), which has some value as a standard of comparison with the real tides. Owing to the rotation of the earth, the tidal force at any point in the sea or air continually changes; thus the tide is a dynamical phenomenon.

During the 18th century and later, Newton's successors developed his dynamical astronomy of the solar system, and succeeded in explaining the planetary motions in almost every detail. They began also to develop the theory of the tides, a task in which Laplace (1799–1830) took a leading part. His *Mécanique Céleste* includes an account of both the planetary theories and the tides. For the most part he restricted his tidal theory to the ideal case of an ocean of uniform depth on a rotating spherical earth. He showed that for such an ocean the tides could be direct or inverse, that is, either high or low tide could occur 'under' the tide-producing body, depending on the ocean depth. Such an ocean has an infinity of modes of *free* oscillation, and resonance might occur between certain of them and the *forced* tidal oscillations. Owing to the irregular boundaries and non-uniform depth of the actual oceans, Laplace's tidal theory has very limited application to them.

Newton (1687b) realized that the tidal forces must affect the atmosphere as well as the oceans, but thought that the atmospheric tides would be too small to be detected. The lack of lateral boundaries renders the atmosphere more appropriate than the seas for Laplace's ideal tidal theory, but the compressibility of the air must be taken into account.

Let p, ϱ, T, g denote the pressure, density and temperature (Kelvin) of the air, and the acceleration of gravity, at height z above the ground. Let the suffix zero added to these and other symbols distinguish the values for $z=0$, that is, the ground values. The equation of static equilibrium of the air is

$$d(\ln p)/dz = -1/H \qquad H = p/g\varrho,$$

where ln denotes the natural logarithm. The name *scale height* is used for H, which in general is a function of z. It is the height of the column of overlying air of *uniform* density ϱ needed to give the pressure p at the height z, if the variation of g with height is ignored (though at 100 km g is reduced by 3%).

In a perfect gas

$$p = knT = R_0 \varrho T/M = R\varrho T,$$

where k denotes Boltzmann's constant (1.38×10^{-16} ergs per °C), n the number density of molecules (per cm³), m the mean molecular mass, and $R_0 = kN$, where N denotes the number 6.02×10^{23} called after Loschmidt (or, less appropriately, after Avogadro). Hence $R_0 = 8.31 \times 10^7$; it is the gas constant per mole, and $R(=R_0/M)$ is the gas constant; $M(=Nm)$ is the mean (chemical) molecular weight of the air (about 29).

Hence

$$H = kT/mg = R_0 T/Mg.$$

Laplace (1799a) considered an atmosphere in which T, m and g, hence also H, are the same at all heights. Then

$$p/p_0 = \varrho/\varrho_0 = e^{-z/H}.$$

He showed that in such an atmosphere the tidal oscillations could be inferred from those for a liquid ocean of uniform depth H (ignoring the mutual gravitation of the water); hence H in this case is called the (tidally) equivalent depth of the atmosphere. In his calculations he took the changes of pressure and density to be isothermal, as Newton had done in his calculation of the speed c of sound waves in air. This led Newton to an erroneous value, $c^2 = p/\varrho = g H$. Laplace corrected this to $c^2 = \gamma p/\varrho$, where γ denotes the ratio of the specific heat of air at constant pressure to that at constant volume; this factor allowed for the *adiabatic* changes of air temperature in such rapid pressure and density changes. Newton's assumption that the airtidal pressure changes are isothermal was natural for such slow changes (but see Section 2L.14).

Laplace's theoretical calculation indicated a 'direct' lunar atmospheric tide with a barometric pressure range in the tropics of $\frac{1}{2}$ mm of mercury. He considered that such a change, though small, should be determinable from a considerable number of hourly readings, but in 1823, when his interest in the lunar air tide was active, no long series of tropical barometric readings was available to him. Instead he used an 8-year series (1815–23) of barometric readings supplied by Bouvard, of the Paris observatory, made there four times daily, at 9, 12, 15 and 21 hours. He used only part of the three daytime readings, 4752 in all; from these, by a well-designed method, using *differences* between successive readings, he sought to compute the lunar semi-diurnal tide. His result for its range was 0.054 mm, with maxima at 3 h 19 m after upper and lower lunar transit. Actually both this and his theoretical estimate were at least four times too big.

Laplace (1825), like De Moivre before him, was a pioneer in error theory, and he stressed the need, in deriving results from observations, to determine also the probability of their being correct within calculated limits; without doing this, he remarked, one risks presenting the effects of irregular causes as laws of nature, 'as has often happened in meteorology'. Having calculated the probability that his result for the Paris air tide was not due merely to chance, that is, to the continual irregular 'weather' changes, he attached limited significance to his determination, saying that at least 40 000 observations would be needed to determine so small a barometric variation with adequate certainty.

Laplace knew that there is a solar daily barometric variation, with a pronounced semidiurnal component, and that this is much larger than his estimate of the lunar semidiurnal air tide. Hence he supposed that it is due mainly to the sun's thermal action. He seems to have thought that there was little hope of constructing a theory of such a thermally excited atmospheric oscillation.

1.2. The Barometric and Other Daily Variations

The sea tides are measured by means of tide gauges, that record the changing height of the water surface. Obviously the air tides cannot be measured in this way, as the atmosphere has no such boundary surface. The alternative is to use a pressure gauge on the bed of the aerial ocean, namely the barometer (the same principle is the only practicable one for the sea tides in mid-ocean). The vertical accelerations of the air are so small that the barometer effectively measures the weight of the overlying air; thus an above-normal barometric height implies a heaping up of air above the station, similar to the heaping up of water at high sea tide. By analogy with the sea tides, the lunar atmospheric tide must cause a rise and fall of the barometer, of lunar semidiurnal period. In the tropics the barometer does show a marked semidiurnal variation, but its period is half a *solar*, not lunar, day; this is illustrated by Figure 1.1 for five days of November 1919, at Batavia (now Djakarta) in Indonesia, at 6°S latitude, and also at the temperate-zone station Potsdam (52.4°N), where the barometer undergoes large irregular variations, associated with weather changes. At Potsdam the semidiurnal variation is not evident, though present.

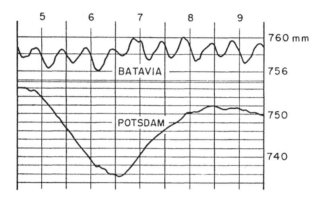

Fig. 1.1.   Barometric variations (on twofold different scales) at Batavia (6°S) and Potsdam (52°N) during November 1919. After Bartels (1928).

It is convenient to denote a solar daily variation, for example of barometric pressure p or atmospheric temperature T (Kelvin), by S – or by $S(p)$, $S(T)$ when the element concerned is to be indicated. Similarly a lunar daily variation is denoted by L. Such daily variations can be analyzed into their harmonic components, with amplitudes S_n, l_n, phases σ_n, λ_n, and harmonic coefficients A_n, B_n, a_n, b_n

$$S = \sum_n S_n, \qquad L = \sum_n L_n,$$

where

$$S_n = s_n \sin(nt + \sigma_n) = A_n \cos nt + B_n \sin nt,$$

$$L_n = l_n \sin(n\tau + \lambda_n) = a_n \cos n\tau + b_n \sin n\tau.$$

$$A_n = s_n \sin \sigma_n, \quad B_n = s_n \cos \sigma_n, \quad a_n = l_n \sin \lambda_n, \quad b_n = l_n \cos \lambda_n.$$

Here t and τ denote respectively mean solar time and mean lunar time, reckoned in angle at the rate 360° per mean solar or lunar day, from lower transit (that is, in the solar case, midnight).

In the series for S, n is taken to be an integer, and usually S is reasonably well represented by the first four terms $n = 1, 2, 3, 4$. In the lunar case the main harmonic component is the semidiurnal one, $n = 2$, and the other harmonic components in general have non-integral values of n (cf. Section 2L.4A).

In the case of $S(p)$ and $L(p)$, the unit in which s_n, l_n, A_n, B_n, a_n and b_n are expressed is usually either the millibar (mb), microbar (μb) or 1 mm of mercury; 1 mb = 1000 μb; and 1 mb = 0.750 mm. In this review these units are generally used from this point onwards, also in quotations where originally other units were used.

1.2A. True or apparent time, and mean time

The true or *apparent* local solar time t' at any station P can be expressed in angular measure, at the rate $15°$ per hour or $360°$ per day, or in hours; in angle it is measured eastward from the true midnight meridian (opposite to the noon meridian, or meridian half-plane bounded by the earth's rotation axis, in which the sun lies) to the meridian of P. For stations on the meridian of Greenwich (G) the time is called Greenwich apparent time t'_u. If ϕ is the longitude of P east of Greenwich, then in angular measure

$$t' = t'_u + \phi.$$

Owing to the varying rate of orbital motion of the earth, the solar day, or interval between two meridian passages of the sun, varies in duration in the course of the year. Hence the hours of apparent solar time are not constant. *Mean* solar time replaces this variable reckoning by a uniform one, with the same zero-point at the vernal equinox (the epoch when the sun is crossing the plane of the earth's equator from south to north), and the same total measure in the course of a year. The difference between Greenwich or local mean time, denoted by t_u and t, and the corresponding apparent times, t'_u and t', is called the *equation of time*, e, so that

$$t = t_u + \phi, \quad t_u = t'_u + e, \quad t = t' + e.$$

The values of e for the mean epoch in each calendar month are as follows:

Jan. $+ 2°19'$	Apr. $+ 0° 4'$	July $+ 1°21'$	Oct. $- 3°28'$
Feb. $+ 3°29'$	May $- 0°52'$	Aug. $+ 0°59'$	Nov. $- 3°47'$
Mar. $+ 2°12'$	June $+ 0° 4'$	Sept. $- 1°11'$	Dec. $- 1° 6'$.

If it is wished to express the series for S relative to apparent solar time t', the phase angles σ_n must be changed to σ'_n, where

$$\sigma'_n = \sigma_n + ne,$$

and A_n, B_n are changed to A'_n and B'_n, given by $A'_n = s_n \sin \sigma'_n$, $B'_n = s_n \cos \sigma'_n$.

The interval of 24 hours of mean solar time succeeding each epoch of midnight at Greenwich is called a Greenwich mean day, or day of universal time (UT). At any station P the local mean day extends from its local mean midnight.

Mean solar time may be associated with a fictitious mean sun that rotates uniformly round the earth; likewise mean lunar time is introduced, associated with the motion of a fictitious mean moon that rotates uniformly round the earth, at the same average rate as the actual moon (cf. Section 2L.4).

1.2B. THE HARMONIC DIAL

A solar daily harmonic component S_n can be illustrated on a 24-hour time base by the graph of $s_n \sin(nt + \sigma_n)$; its range is $2s_n$ and its n maxima and n intervening minima are spaced at $24/2n$ hour intervals; the time of first maximum is the hour $t_n = (90° - \sigma_n)/15°n + 24 \, r/n$, where r is the least integer that makes t_n positive and less than $24/n$. This is illustrated by the part (a) of Figure 1.2 (Bartels, 1932a) for $S_2(p)$ at Washington,

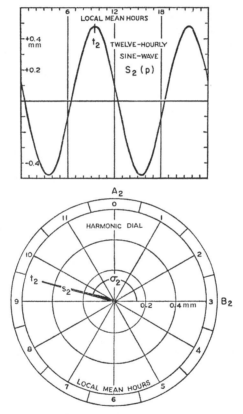

Fig. 1.2. The solar semidiurnal variation of barometric pressure, $S_2(p)$, at Washington, D.C., represented (above) by a daygraph, and (below) by a harmonic dial. After Bartels (1932a).

D.C. An alternative way of *specifying* S_n graphically is to show its amplitude s_n and phase σ_n, also its coefficients A_n, B_n, by a *harmonic dial*. In its full form, as shown for $S_2(p)$ in Figure 1.2(b), the dial has the face of a $24/n$ hour clock, round which the hour hand makes n circuits per day; thus for $n=2$, as in Figure 1.2(b), it shows the usual 12 hours; for $n=1$ it has a 24-hour clock face, and so on. The upward vertical and rightward horizontal directions are marked as A_n and B_n axes, and the point P

with coordinates A_n, B_n (on a scale indicated) is marked and joined to the center O of the dial. The vector OP has the length s_n on the same scale, and it makes the angle σ_n, measured anticlockwise, with the axis OB_n; it also shows the time t_n of first maximum of the graph of S_n. If OP is made to rotate anticlockwise n times daily, its projection on the OA_n axis after time t is $s_n \sin(nt + \sigma_n)$.

Lunar daily harmonic components L_n can be similarly represented by a graph or specified graphically by a dial. For them the time base of the graph is one mean lunar day, which may be divided into 24 lunar hours; and the dial hour hand makes n circuits in a lunar day.

The part S_1 of the daily variation S is called the diurnal component or variation. *Diurnal* is here used always to mean a *harmonic* variation with period one day; we do not use it merely as a synonym of *daily*, as is often done. Likewise the terms S_2, S_3, S_4 are respectively called the semidiurnal, terdiurnal and quaterdiurnal components or variations. We adopt a similar usage in the case of the lunar daily variation L and its harmonic components L_1, L_2,..., which we call respectively lunar diurnal, semidiurnal,... components or variations, even though in this case some values of n may only approximate to an integer.

It is often convenient and adequate to use the dial representation of the amplitude and phase of a harmonic component, but to omit much of the dial background and notation; also it suffices to show only the end point P of the vector. But the scale and the A_n, B_n (or a_n, b_n) axes must be drawn. This is illustrated by Figure 1.3 (Bartels, 1927), which shows 40 dial points for $L_2(p)$ at Batavia. Each indicates the deter-

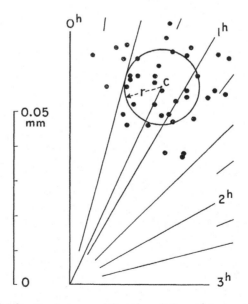

Fig. 1.3. Harmonic dial for the lunar semidiurnal variation of barometric pressure, $L_2(p)$, at Batavia, for each of the years 1866–1905; C, the centroid of the 40 annual dial points, indicates the 40-year mean $L_2(p)$. The circle centered on C is the probable error circle for any one of the annual dial points. After Bartels (1927).

mination for one calendar year of the period 1866 to 1905, so that all the points are of equal weight. The centroid C, the mean position of the 40 points, likewise indicates the mean $L_2(p)$ for the whole 40 years. Bartels (1932b) applied the theory of errors in a plane, on the assumption that the distribution is Gaussian, to assess the uncertainty of such individual results and that of their mean. He showed how to determine the *probable error ellipse*, which should include half of the points (see also Chapman and Bartels, 1940, Section 19.2). Where, as in Figure 1.3, the individual points are approximately symmetrically distributed around C, the ellipse becomes the *probable error circle*, there shown. The radius r of the circle is the probable error of any one of the yearly determinations; it is $0.989d$, where d denotes the mean distance of the points from C. The probable error r_0 of the mean of n such determinations, all, as here, of equal weight, is $r/(n-1)^{1/2}$.

A mean determination may be considered reasonably good if l_2 is at least equal to three times its probable error r_0 (the determination has significance though less accuracy even if l_2/r_0 is only 2). In Figure 1.3, $l_2 = 0.062$ mm, and r for the yearly points is 0.011 mm; thus $r_0 = 0.0016$, and l_2/r_0 is 39, indicating a high accuracy for the 40-year result.

The uncertainty of the mean phase may be regarded as corresponding to the angle $\sin^{-1}(r_0/l_2)$, half the angle subtended at O by the circle of radius r_0 centered on C. Thus in Figure 1.3 the phase uncertainty for the mean determination is $\pm 1.5°$, or ± 6 min of time. This does not represent a probable error, and somewhat exaggerates the uncertainty of the phase. This is because within the angle subtended at the origin by the probable error circle there are points outside the circle, on the near and further sides of the circle. Neither the amplitude nor the phase has a Gaussian distribution.

The variation $S_2(p)$ (shown for Batavia in Figure 1.1 and for Washington, D.C. in Figure 1.2) is one of the most regular of all meteorological phenomena. It is readily detectable by harmonic analysis also at stations like Potsdam, where it is overlaid by much larger irregular variations (Figure 1.1), whose range within a week or less may be 20 mm of mercury. Tropical barographs, on the other hand, show a decidedly regular semidiurnal variation with a daily range of about 2 mm, and a weekly range not much greater, except during hurricanes. Humboldt carried a barometer with him on his famous South American journeys of 1799–1804. In his book *Cosmos* he remarked that the two daily maxima at about 10 a.m. and 10 p.m. were so regular that his barometer could serve somewhat as a clock.

This regularity is well illustrated by Figure 1.4 (Bartels, 1932a), which shows dial vectors for the annual mean $S_2(p)$ for many places in North America. The vectors relate to the map points at the thick end of each vector. The phases are all similar, and the amplitudes decrease rather regularly from south to north. However, as was noted by the Austrian meteorologist Hann, who gave much attention to the distribution and seasonal variations of $S_2(p)$ over the globe, the amplitude s_2 is reduced near the Pacific coast. Hann found likewise that s_2 is less on the east Adriatic coast than in Italy, and in the West Indies compared with the East Indies (now Indonesia).

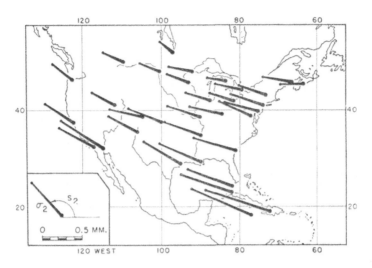

Fig. 1.4. The distribution of the harmonic dial vectors for $S_2(p)$ at many stations over the U.S.A. Each dial vector indicates $S_2(p)$, on the scale shown by the inset, for the point at the thick end of the vector. After Bartels (1932a).

1.3. Thermal Tides and Kelvin's Resonance Theory

Thomson (later known as Lord Kelvin), in a presidential address to the Royal Society of Edinburgh in 1882, discussed the harmonic components of the solar daily variations of atmospheric pressure and temperature as follows, in connection with a table of values he quoted, that gave $S_1(p)$, $S_2(p)$ and $S_3(p)$ for thirty different places:

The cause of the semi-diurnal variation of barometric pressure cannot be the gravitational tide-generating influence of the sun, because if it were there would be a much larger lunar influence of the same kind, while in reality the lunar barometric tide is insensible, or nearly so. It seems, therefore, certain that the semi-diurnal variation of the barometer is due to temperature. Now, the diurnal term, in the harmonic analysis of the variation of temperature, is undoubtedly much larger in all, or nearly all, places than the semi-diurnal. It is then very remarkable that the semi-diurnal term of the barometric effect of the variation of temperature should be greater, and so much greater as it is, than the diurnal. The explanation probably is to be found by considering the oscillations of the atmosphere, as a whole, in the light of the very formulas which Laplace gave in his *Mécanique céleste* for the ocean, and which he showed to be also applicable to the atmosphere. When thermal influence is substituted for gravitational, in the tide-generating force reckoned for, and when the modes of oscillation corresponding respectively to the diurnal and semi-diurnal terms of the thermal influence are investigated, it will probably be found that the period of free oscillation of the former agrees much less nearly with 24 hours than does that of the latter with 12 hours; and that, therefore, with comparatively small magnitude of the tide-generating force, the resulting tide is greater in the semi-diurnal term than in the diurnal.

In 1890 the 3rd Lord Rayleigh, author of the famous treatise on sound, discussed $S_2(p)$ and $S_1(p)$ in a similar vein, without mentioning Kelvin, as follows:

Beforehand the diurnal variation of the barometer would have been expected to be much more conspicuous than the semi-diurnal. The relative magnitude of the latter, as observed at most parts of the earth's surface, is still a mystery, all the attempted explanations being illusory. It is difficult to see how the operative forces can be mainly semi-diurnal in character; and if the effect is so, the readiest explanation would be in a near coincidence between the natural period and 12 hours. According to this view the semi-diurnal barometric movement should be the same at the sea-level all round the earth, varying (at the equinoxes) merely as the square of the cosine of the latitude, except in consequence of local disturbances due to want of uniformity in the condition of the earth's surface.

In the same paper Rayleigh estimated the free periods of oscillation of the atmosphere, of diurnal and semidiurnal type, to be respectively 23.8 and 13.7 hours. These certainly did not support Kelvin's resonance suggestion, but they were not regarded by Rayleigh as likely to be correct, because he ignored the earth's rotation, and treated the atmosphere as being at uniform temperature, and the pressure and density changes associated with the waves as taking place isothermally. Regarding the latter, he remarked as follows:

In a more elaborate treatment a difficult question would present itself as to whether the heat and cold developed during atmospheric vibrations could be supposed to remain undissipated. It is evidently one thing to make this supposition for sonorous vibrations, and another for vibrations of about 24 hours period. If the dissipation were neither very rapid nor very slow in comparison with diurnal changes – and the latter alternative at least seems improbable – the vibrations would be subject to the damping action discussed by Stokes.

This question was later considered by Chapman (1932a, b), theoretically and observationally (see Section 2L.14).

The lonely ill-fated Austrian meteorologist Margules (1890, 1892, 1893) (Platzman, 1967) studied the free and thermally excited forced oscillations of the atmosphere in much detail, on the basis of Laplace's theory, and with the explicit purpose of testing Kelvin's resonance hypothesis. He concluded that there would be close resonance for $S_2(p)$. But all his calculations were based on atmospheric models now known to differ much from reality. In the most realistic of his calculations of thermally excited oscillations, he took the amplitude of $S_2(T)$ as decreasing exponentially upward, as would be the case for a periodic variation conducted upward from the ground with uniform conductivity. But he did not include the accompanying variation of the phase of $S_2(T)$ with height, as was later pointed out by Chapman (1924a), who incorporated this feature of $S_2(T)$ in his own calculations.

1.4. More Realistic Atmospheric Models

Lamb (1910) made an important extension of Laplace's theory. His work related to an atmosphere on a plane base, thus abstracting from the problem the sphericity and rotation of the earth; but later, in the 4th (1916) edition of his *Hydrodynamics*, he removed these last two restrictions, though only for an atmosphere in convective equilibrium. His main discussion referred to an atmosphere in which H varies uniformly with the height ($H = H_0$ being a special case). He showed that the propagation of long waves in such an atmosphere is similar to that of long waves in a liquid ocean of depth H_0 in two special cases, namely, (1) Laplace's case, in which $H = H_0$ (or $T/m = T_0/m_0$) at all heights, and the density variations occur isothermally; and (2) for an atmosphere in adiabatic equilibrium, and in which the density variations occur adiabatically. The height of such an atmosphere is $\gamma H_0/(\gamma-1)$; the temperature T decreases uniformly upwards to zero at the rate $(\gamma-1)T_0/\gamma H_0$. (Laplace's case can be considered as corresponding to $\gamma=1$.)

The actual atmosphere is not in adiabatic equilibrium, so that it cannot be supposed, at least without further proof, that the tidally equivalent (liquid ocean) depth for the atmosphere is H_0. Lamb in fact showed that when H varies linearly with height, but not adiabatically, there is an infinite series of speeds for long waves, with the implication that there is a similar series of values of the equivalent depth h. However, the impression persisted widely for over twenty years that for any type of atmosphere there is just one value of h.

Lamb briefly discussed the resonance hypothesis of S_2 in his 1910 paper and in subsequent editions of his *Hydrodynamics*. He estimated from the improved form of Laplace's theory given by Hough (1897, 1898), in terms of spherical harmonic functions, that if the atmosphere is resonant with a free oscillation similar to S_2 in its geographical distribution, the equivalent ocean depth must be about 8 km; whereas for the actual atmosphere H_0 varies from about 7.3 km at the poles to 8.7 km at the equator. He continued:

Without pressing too far conclusions based on the hypothesis of an atmosphere uniform over the earth, and approximately in convective equilibrium, we may, I think, at least assert the existence of a free oscillation of the earth's atmosphere, of 'semi-diurnal' type, with a period not very different from, but probably somewhat less than, 12 mean solar hours.

He continued further:

At the same time, the reason for rejecting the explanation of the semi-diurnal barometric variation as due to a gravitational solar tide seems to call for a little further examination. The amplitude of this variation at places on the equator is given by Kelvin as 1.08 mb. The amplitude given by the 'equilibrium' theory of the tides is about 0.016 mb. Some numerical results given by Hough in illustration of the kinetic theory of oceanic tides indicate that in order that this amplitude should be increased by dynamical action some seventy-fold, the free period must differ from the imposed period of 12 solar hours by not more than 2 or 3 minutes. Since the difference between the lunar and solar semi-diurnal periods amounts to 26 minutes, it is quite conceivable that the solar influence might in this way be rendered much more effective than the lunar. The real difficulty, so far as this point is concerned, is the *a priori* improbability of so very close an agreement between the two periods. The most decisive evidence, however, appears to be furnished by the phase of the observed semi-diurnal inequality, which is accelerated instead of retarded (as it would be by tidal friction) relatively to the sun's transit.

1.5. The Phase of $S_2(p)$

Chapman (1924a) stressed the argument for strong resonance of S_2, based on the regularity of its geographical distribution, as compared with the considerable non-uniformity of the solar semidiurnal variation of air temperature (especially as between land and sea areas), which according to Lamb's last-quoted remark must be at least an important part of the cause of S_2. This argument he strengthened by contrasting the regularity of the geographical distribution of S_2, due partly to an irregular cause, with the degree of irregularity shown by $L_2(p)$, whose cause is certainly distributed very regularly.

Chapman also extended Margules' calculation of the oscillations produced by the semidiurnal component of the daily variation of air temperature $S_2(T)$, taking account of the variation of phase, later discussed, in this connection, by Bjerknes (1948), as well as of amplitude, with height. He concluded that the phase of the part of $S_2(p)$ that is of thermal origin must be about 135° in advance of the phase of $S_2(T)$, which he tried to estimate from the temperature data collected by Hann (1906), Brooks (1917) and others, also using the Taylor (1917) estimate of the thermal conductivity due to eddy motion.

Chapman also compared the magnitudes of the thermal and tidal contributions to $S_2(p)$, which is possible according to his assumption that (substantially) both are affected by the same resonance magnification. He inferred that they are of roughly equal order of magnitude (the inadequacy of the data regarding $S_2(T)$ precluded a more accurate statement). On this basis he could explain the observed phase of $S_2(p)$ from the phase of the thermal part, inferred from $S_2(T)$, and on the assumption that the tidal part is in phase with the sun. This further enabled him to estimate the factor of resonance magnification as about 100. He was unable to prove that the atmosphere has a free period of oscillation (of the right geographical distribution) which would give this magnification. As the resonance magnification (when considerable) would be proportional to $1/(t_i - t_f)$, where t_i denotes the imposed period and t_f the free period, he concluded that despite the *a priori* improbability, $t_i - t_f$ cannot exceed 2 or 3 min, and is positive.

1.6. Doubts as to the Resonance Theory

Whipple (1918, also in 1924, in the discussion on Chapman's paper), found great difficulty in accepting the resonance theory, on account of the possibility that such accurate 'tuning' of the forced to the free oscillation might be upset by the large changes in air pressure and temperature associated with weather and annual variations, and also on account of the difficulty the S_2 wave would have in twice daily surmounting the heights of Central Asia and the Rocky Mountains without losing a material fraction of its energy. These irregularities, however, are on a relatively small scale. The possibility that the variations of mean temperature from year to year might affect the tuning was examined by Bartels (1927), but no such effect was found.

Taylor (1932) also was led to doubt the validity of the resonance hypothesis, despite the strength of the general arguments in its favour. Lamb had assumed, on the basis of the work already mentioned, that free oscillations of the atmosphere can exist which are identical, in distribution and period, with those of an ocean of such depth that long waves are propagated in it with the speed that he calculated for plane atmospheric waves. This assumption was used by Taylor (1929, 1930) to estimate the period of the oscillation of S_2-type to be expected in an atmosphere in which the speed of propagation of long waves was that of the waves produced by the Krakatoa volcanic eruption of 1883. This great atmospheric pulse was propagated more than once round the entire earth, with a speed of 319 m sec^{-1}, which corresponds to an equivalent depth $h = 10.4$ km, a value markedly too great to give a free period (for an oscillation of S_2-type) nearly equal to 12 hours.

Lamb's assumption may be regarded as an extension of Laplace's theory of waves in an isothermal atmosphere. As realized later by Taylor, it involves the possibility that the atmosphere may have many equivalent depths, a contingency not possible in the atmospheres of the special type considered by Laplace and Lamb, for which $h = H_0$. Lamb's assumption is not obviously true, but later Taylor (1936) proved its validity, and further developed Lamb's investigation of oscillations that are distributed in a similar way geographically (that is, as functions of longitude ϕ and colatitude θ), but have different height distributions of motion. In this work he was the first to take account of the cessation, at the tropopause, of the upward decrease of temperature.

1.7. Renewed Hope in the Resonance Theory

Pekeris (1937) applied Taylor's methods to determine the free periods of an atmosphere in which the stratospheric temperature increases upwards above a certain height. This temperature distribution had been inferred from studies of the abnormal propagation of sound to great distances (beyond a zone of silence surrounding the source of sound), as well as from the heights of appearance and disappearance of meteors. Pekeris showed that, subject to a certain condition, the atmosphere could oscillate in ways corresponding to two equivalent oceanic depths. One of these was about 10 km, associated with a speed of propagation equal to that of the Krakatoa wave; the other gave a period (for a geographical distribution of S_2-type) of very nearly 12 hours, though the uncertainty of the upper atmospheric data precluded an exact calculation of the free period. The condition referred to was that the atmospheric temperature, after increasing upward above the stratosphere, should reach a maximum and thereafter decrease upwards to a low value. This was in agreement with the temperature distribution proposed by Martyn and Pulley (1936).

An important conclusion reached by Pekeris on this basis was that at high levels the daily air motions may be reversed in phase and highly magnified. This fitted well with the dynamo theory of the lunar daily magnetic variation, the phase of which is about opposite to that inferred from a simple consideration of $L_2(p)$ at the ground, and in the light of the present knowledge of the electrical conductivity of the ionosphere. A large amplitude of the lunar daily air motions is also suggested by the lunar tidal change of height of the E layer of the ionosphere, found by Appleton and Weekes (1939), over London, England, although the phase of this variation does not show the expected reversal.

Pekeris (1939) re-examined the barometric traces for the Krakatoa waves, and found evidence of a minor component wave propagated with the speed of 250 m sec^{-1}, corresponding to an equivalent depth of 7.9 km, which would accord with a free oscillation (of S_2-type) with a period nearly equal to 12 hours. He showed that as the explosion occurred at a low level, most of the energy of the pulse should go into the faster-travelling wave. (He estimated this energy as about 10^{24} ergs, roughly 1000 times that of the waves set up by the 1908 Siberian meteorite. Whipple (1930) had examined the waves set up by that meteorite.)

1.8. Atmospheric Oscillations as Studied by Weekes and Wilkes

Weekes and Wilkes (1947) used the methods given by Pekeris (1937) to study the atmospheric oscillations, especially those of semidiurnal type. They made many numerical calculations of the free period of oscillations of this type, for different tentative models of the height-distribution $T(z)$ of temperature T with height z. For this purpose they used a then modern differential analyzer, which gave results much more speedily than had previously been possible. Their attitude was one of such confidence in the resonance theory that it was used to infer properties of the atmosphere beyond the heights for which observations were then available. This was before rocket measures of T had been published. (For a more recent attempt to deduce T from tidal data, cf. Section 3.5.A).

Wilkes (1949), in his monograph on atmospheric oscillations, reviewed the history of the subject since the time of Laplace, and gave the derivation of Laplace's tidal equation, the starting point for all such computations. Laplace obtained it while discussing the free oscillation, under gravity, of a uniform liquid ocean of uniform depth h on a spherical earth of radius a rotating with angular velocity ω. The solution has naturally to satisfy certain boundary conditions. In order to describe the work of Weekes and Wilkes, we must first present a brief treatment of tidal theory. The subject is dealt with in detail in Chapter 3.

The independent variables in the problem are the time t, and the coordinates of any point:

$$\theta \text{ the colatitude, } \phi \text{ the east longitude, } z \text{ the height above the ground.} \quad (1)$$

The five equations of motion and of state determine five dependent variables, which may be chosen in more than one way, from among the following: the components of the velocity $\mathbf{V}$,

$$u \text{ southward,} \quad v \text{ eastward,} \quad w \text{ upward,} \quad (2)$$

the departures of p, ϱ, T,

$$\delta p, \quad \delta \varrho, \quad \delta T \quad (3)$$

from their mean values at the point: the divergence χ of the velocity, div $\mathbf{V}$ or G, thus defined:

$$G = -\frac{1}{\gamma p_0} \frac{\mathrm{D}p}{\mathrm{D}t}, \quad (4)$$

where $\mathrm{D}/\mathrm{D}t$ denotes the 'mobile operator' $\partial/\partial t + (\mathbf{V} \cdot \mathbf{grad})$.

Normal modes of free oscillation are sought, in which each dependent variable is proportional to

$$\exp i \left(s\phi + \sigma t \right), \tag{5}$$

where s denotes an integer; thus the free period is

$$2\pi/\sigma. \tag{6}$$

The dependence on z and θ is taken to be separable, the latter being expressed by a function $\Theta(\theta)$. Laplace's tidal equation is a differential equation to determine Θ, and has the form:

$$\left(F + 4a^2\omega^2/gh \right) \Theta \left(\theta \right) = 0, \tag{7}$$

where F denotes a differential operator (given later, in Section 3.2), containing f, given by

$$f = \sigma/2\omega, \tag{8}$$

as a parameter. For each f there is a corresponding function Θ.

Laplace's tidal equation is applicable to the *forced* oscillations of an atmosphere, whatever its height distribution $T(z)$ of temperature T (supposed uniform over the globe). In this case σ is known, as well as s, namely $2\pi/\sigma$ is the imposed period. The tidal equation in this case determines a series of values of Θ and of h, whose original oceanic definition has no application to an atmosphere without any upper bound.

The height distribution of the five chosen dependent variables in this case has to be determined by a separate differential equation. This may be conveniently expressed in terms of the pressure at each height, as an independent variable, by taking $x = -\ln \{p_0(z)/p_0\}$. If we write

$$\operatorname{div} \mathbf{V} = -\operatorname{D} \ln \left(\rho_0 + \delta\rho \right) / \mathrm{D}t = \{(p_0(z)/p_0)\}^{\frac{1}{2}} y, \tag{9}$$

the equation is

$$\frac{\mathrm{d}^2 y}{\mathrm{d}x^2} + \left[-\frac{1}{4} + \frac{1}{h}\left(\frac{\gamma-1}{\gamma} H + \frac{\mathrm{d}H}{\mathrm{d}x} \right) \right] y = 0. \tag{10}$$

Here the height-distribution of the atmosphere (depending on the temperature T and the mean molecular mass m) is involved through the scale height H. In so far as y can be considered as of fairly constant order of magnitude (and this is a matter for examination by means of this equation), the above expression for $\operatorname{div} \mathbf{V}$ indicates an upward increase of $\operatorname{div} \mathbf{V}$ inversely proportional to $p^{1/2}$, that is, by 1000-fold at about the height of the E-layer.

Weekes and Wilkes (1947) gave an interesting interpretation of this equation by analogy with the propagation of electromagnetic waves in a medium having a variable refractive index. In the atmospheric case the expression for the analogous 'equivalent' refractive index is

$$\mu^2 = -\frac{1}{4} + \frac{1}{h}\left(\frac{\gamma-1}{\gamma} H + \frac{\mathrm{d}H}{\mathrm{d}x} \right). \tag{11}$$

If the height-distribution of H, for any value of h, makes μ^2 negative at a certain level, upward propagation of the energy, if mainly put into the atmosphere by tidal or thermal causes in the lower layers, is effectively blocked. The air at that height acts as a barrier, total or partial, trapping the energy, and building up the amplitude in the whole spherical shell between the ground and the barrier, giving rise to resonance. If μ^2 is negative, not for all heights above the level at which it first becomes zero, but only for an interval of height, the barrier is partially transparent, and some of the oscillatory energy passes through it, either to a second (or third) barrier where there is a height interval of negative μ^2, or to the high levels at which thermal conductivity and dissipation of the energy into heat by viscosity become important. At these high levels the condition that $\delta p/p$ or $\delta \varrho/\varrho$ is small, as assumed in the equation, may cease to hold, and the modified differential equations can become nonlinear.

The conditions favouring negative μ^2 are that H should be small and that $\mathrm{d}H/\mathrm{d}x$ should be either positive and small, or negative, corresponding to an upward decrease of temperature, because x increases upwards.* The number of barriers to energy flow depends on the number of such regions of upward-decreasing temperature, but they alone are not sufficient to give a barrier, unless the value of h is appropriate, which in turn depends on the mode and period of the oscillation under consideration.

The boundary conditions in the equation for y are that at high levels the energy flow is upwards, and that at the ground ($z=0$ and $x=0$) the vertical velocity is zero – unless we are taking into account the tidal motion of the seas at the base of the atmosphere, or the varying level of the ground (cf. Section 3.6A).

Weekes and Wilkes (1947) had only indirect information about $T(z)$ for the earth's atmosphere, and made calculations of the free periods for a variety of distributions. Wilkes (1949), in his monograph, was able to quote the distribution of $T(z)$ given by Best, Havens, and La Gow (1947), determined from one of the earliest rockets used for upper atmospheric research. It agreed extremely well with what then seemed indicated by all the indirect evidence except that of meteor trails (whose interpretation was uncertain). Wilkes (1949, p. 64) concluded thus:

The resonance theory may now be taken as well established;

he also countered one of Whipple's criticisms thus:

It is now realized that the large variations of atmospheric temperature and pressure associated with weather changes are very localized in nature, and affect only the first 15 km or so of the atmosphere, whereas the solar semi-diurnal oscillation is world-wide in extent and embraces the atmosphere up to the height of the E region. The vast scale of the oscillation also provides the answer to another question put by Whipple, namely how it is possible for the wave to surmount the height of the Rocky Mountains at each revolution, without losing an appreciable proportion of its energy.

Later, discussing thermally excited oscillations, Wilkes (1951) used the methods of Pekeris (1937) to calculate the oscillation excited by temperature variations due to heat conducted upward from the ground. He gave some numerical results, based on

---

* In addition, it is possible, as is shown in Section 3.5, for h to be negative.

estimated values of $S_2(T)$ over the globe; he "found them not to conflict with the hypothesis put forward by Chapman in 1924, that the thermal and gravitational contributions to this oscillation are of the same order of magnitude". In the same paper he briefly wrote on the dynamical effect of periodic heating and cooling in a layer not in contact with the ground, but said that: "Since this case is not of great practical interest, the expression of the equivalent gravitational oscillation" would not be given.

In the following year, in a review lecture, Wilkes (1952) was more guarded as to the reliability of his calculation of the relative magnitudes of the thermal and gravitational shares in the excitation of $S_2(p)$; he said that owing to the uncertain value of $S_2(T)$ as an average over the globe, it is not to be expected that estimates of the ratio will be very accurate; but that "if the balance had been overwhelmingly in favor of thermal action, the calculations should have revealed it. This they did not do, and the question must remain one for further speculation."

Chapman (1951), in a review article*, described the studies of the resonance theory up to that time. Its status was indicated thus:

The existing theory [namely, the resonance hypothesis] will need to be revised as our knowledge of the upper atmosphere advances through rocket investigations and in other ways ...,

1.9. Rockets Exclude Resonance

Jacchia and Kopal (1952) followed Wilkes' methods in calculating numerically the amplification of the sun's gravitational tide for several atmospheric models, with the objective "to account for the observed magnification of the surface pressure oscillations with periods of 12 and 10.5 hours". Among the height distributions of temperature (or profiles) $T(z)$, they mentioned specially the NACA (National Advisory Committee for Aeronautics) standard atmosphere (Diehl, 1948), and one based on a smoothed mean of the temperatures derived from pressure gauges launched from White Sands, New Mexico, U.S.A., on V-2 rockets, and of the temperatures obtained from microbarographic records of the Heligoland explosion. Neither of these model atmospheres gave adequate magnification, thus, their conclusions on the basis of presumably the best temperature profiles then available were inimical to the resonance theory. But "in the search for the temperature profile which would yield the greatest amplification factor for the semidiurnal oscillations", they were successful in obtaining a profile that for a forced period of 11h58m gave a factor 81 (and another large factor for a period of 10h55m). They concluded: "This profile may therefore be regarded as satisfactory from the dynamical point of view, and its consequences are in sufficiently close agreement with the observational evidence..." (p. 22), and that this profile "agrees with the direct temperature measurements in the lower atmosphere, is consistent with the results pertaining to the upper atmosphere

---

* Permission to quote here extensively from that review has been given by its publisher, the American Meteorological Society, and its sponsor, the Air Force Cambridge Research Center.

within the limits of their uncertainty, and, in addition, leads to sufficiently large pressure oscillations with periods of 12 and 10.8 hours" (p. 13).

However, later rocket measurements of $T(z)$ did not agree with the NACA atmosphere, nor with the tentative Jacchia-Kopal profile, both of which, like the Martyn and Pulley profile discussed by Pekeris, and the one by Best *et al.* (1947) quoted by Wilkes (1949), gave too high a temperature (320° to 350°) at the stratopause at about 50 to 60 km height. Jacchia and Kopal found that the calculated surface magnification of the solar gravitational tide is very sensitive to the value of the temperature at this level. The later rocket determinations of $T(z)$ have much reduced the estimate of T at 50 km. One of the most recent authoritative Standard Atmospheres available (CIRA, 1965) gives the value as 271 K. The rocket results definitely exclude the possibility of notable magnification of $S_2(p)$ by resonance; they give the deathblow to Kelvin's hypothesis.* Chapman's explanation of the phase of $S_2(p)$ likewise ceases to be admissible. The gravitational part of $S_2(p)$ is likely to be similar to or less than $L_2(p)$, and consequently $S_2(p)$ must be produced almost entirely thermally.

Another speculation by Kelvin that must be abandoned (Siebert, 1961, p. 177), and of course also its development by Holmberg (1952), is that the angular velocity of the earth may be regulated by the torque exerted by the gravitational attraction of the sun on the oblateness of the atmosphere, associated with $S_2(p)$. But as late as 1955 Haurwitz and Möller concluded that:

Holmberg's hypothesis combined with the observational evidence on atmospheric tides and with our knowledge of the stratification of the atmosphere make it highly probable that the resonance theory provides the correct explanation of the atmospheric tides.

There are other worldwide atmospheric components of the daily variation $S(p)$ of pressure, which have been studied by various writers, especially $S_3(p)$ by Hann (1918), and $S_4(p)$ by Pramanik (1926). More recently Siebert (1957) and Kertz (1959) used these oscillations, whose cause must be thermal, to study empirically the resonant properties of the atmosphere. No clear indication of resonance could be gained from them.

1.10. Ozone Absorption of Radiation the Main Cause of $S_2(p)$

Thus the rational theory of $S_2(p)$ proposed by Kelvin has proved to be false, like that of Bacon for the sea tides. But the theory of atmospheric oscillations has been greatly advanced by the long series of studies undertaken to confirm or disprove Kelvin's hypothesis. He suggested it to explain the contrast between the predominance of $S_2(p)$ over $S_1(p)$ at the surface, as against that of $S_1(T)$ over $S_2(T)$ at the surface, assuming that the main cause of $S(p)$ is thermal. His idea being no longer tenable, that the contrast arises from a selective magnification by resonance of $S_2(p)$, one may conclude as an alternative (Siebert, 1961, p. 114) that $S_1(p)$ is suppressed, at

* A more careful inclusion of the effects of known sources of dissipation in the atmosphere would likewise have shown the impossibility of the resonance theory (cf. Section 3.6B).

least at ground level. This idea can be illustrated by the oscillation of a chain, one point of which is given a forced displacement with period T. Even though the first harmonic component of this displacement, with period T, has an amplitude much exceeding that of any of its higher harmonic components, the latter may have greater amplitudes near nodes of the T harmonic elsewhere along the chain. The effect need not be particularly profound, since the global average of $S_1(p)$ at the ground is about half the amplitude of $S_2(p)$ (Haurwitz, 1965). First, however, one must account for $S_2(p)$.

Margules, Chapman, Pekeris and others, in considering the thermal excitation of $S_2(p)$, all took account only of the temperature changes caused by upward eddy conduction from the ground. The ground was tacitly supposed to be the only effective absorber of the solar radiation not intercepted and reflected outward by clouds. But some of the radiation does not get to the ground; it is absorbed during its passage through the air, and this affects the air at all levels, not only near the ground. The need to consider the effect of such heating at all levels was indicated by Siebert (1954) and by Sen and White (1955). Siebert (1954, 1956a) suggested that absorption by water vapor might explain or significantly contribute to the explanation of $S_2(p)$. In 1961 he gave the first numerical estimate (Siebert, 1961, Section 7); though the water vapor absorption makes only a small change, $S(T)$, this decreases upward more slowly than the part of $S(T)$ due to conduction from the ground. The result is that the water vapor absorption is about ten times as effective for $S_2(p)$ as the upward conduction. However, the combined effect is only about a third of $S_2(p)$. Siebert (1961) examined ozone absorption, at a higher level, as one of the possible ways of accounting for the main part of $S_2(p)$, but found its effect to be small. In his calculations, however, he assumed the atmosphere above the tropopause to be isothermal and very cold (160 K). As a result H was very small and μ^2 was negative (see Equation (1.11)) above the tropopause; consequently a semidiurnal oscillation excited near 50 km could not effectively propagate to the ground. Butler and Small (1963), using a more realistic distribution of T, found that ozone heating could, indeed, account for about two thirds of the observed $S_2(p)$ at the ground. Although Butler and Small showed that the atmospheric temperature distribution influenced $S_2(p)$, Lindzen (1968a) showed that the surface pressure oscillation is quite insensitive to details of the atmospheric thermal structure – as long as the mean temperature is not too unrealistic. Thus, between water vapor and ozone absorption, $S_2(p)$ at the ground seems roughly accounted for. The present explanation of $S_2(p)$ is a nonresonant explanation in the sense that it does not call for a highly tuned atmosphere. Butler and Small also showed that ozone absorption could substantially account for $S_3(p)$.

Having accounted for $S_2(p)$, we are left, as Siebert (1961) suggested, with the problem of explaining why $S_1(p)$ is suppressed at the ground. Butler and Small (1963) suggested that the main diurnal mode has a short vertical wavelength, while ozone heating is distributed over a great depth of the atmosphere. Hence diurnal disturbances would be reduced by destructive interference. As we shall see, this explanation is incomplete.

The whole question of S_1 has become more important with the advent of upper air data from balloons, rockets, radar tracking of meteor trails, and, indirectly, from geomagnetic dynamo calculations. These data show that S_1 is comparable with, and often much larger than, S_2 in the atmosphere above the ground. Clearly while S_1 may be suppressed at the ground, this is not the case in much of the atmosphere.

1.11. Upper Air Data

Given the relatively small amount of data available for upper atmospheric fields (most notably wind), it may be difficult to see how tidal and thermotidal components can be reliably isolated. However, as we see from measurements of $S_2(p)$ in the tropics, where the thermotidal oscillation is large and other sources of variation are small, then a short data series suffices for a meaningful determination of amplitude and phase. This is frequently the case in the upper atmosphere, where thermotidal oscillations are major components of the total meteorological variation. In part, this is because the amplitudes of atmospheric oscillations tend, under certain circumstances, to grow as $\varrho_0^{-1/2}$ (see Section 1.8). A more detailed discussion of upper air data and the problems involved in its analysis is presented in Chapter 2. What follows is a brief sketch of the findings.

Using several years of radiosonde data for stations between 30°N and 76°N, Johnson (1955), Harris (1959) and Harris, Finger and Teweles (1962, 1966) have determined solar diurnal and semidiurnal contributions to temperature and horizontal wind fields between the ground and 30 km. Typical amplitudes for oscillations in the horizontal wind are 20 cm/sec in the first few kilometers of the atmosphere and 45 cm/sec near 25 km. No pronounced disparity between the diurnal and semidiurnal contributions was found. Wallace and Hartranft (1969) found that at temperate and tropical latitudes, diurnal wind oscillations are significantly associated with local topographic features at the earth's surface. However, at higher altitudes (above 25 km) and at arctic latitudes the diurnal oscillations are global – following the sun.

Over the last ten years the Meteorological Rocket Network has produced a substantial body of data on winds between 40 and 60 km, mostly at White Sands Missile Range (30°). Analysis of this material by Miers (1965), Beyers, Miers and Reed (1966), Reed, McKenzie and Vyverberg (1966a, 1966b) and Reed, Oard and Sieminski (1969) shows the existence of a strong diurnal wind oscillation with an amplitude of about 8 m/s at 50 km. At these altitudes the diurnal oscillation appears to be the main component of the total north–south wind, and there appears to be little evidence of a comparable semidiurnal oscillation (Reed, 1967).

In the altitude range 80–100 km an important source of wind information comes from the ionized trails left by the numerous meteors disintegrating there. These trails are carried by the neutral wind and may be tracked from the ground by the observation of reflected radio signals. Using such techniques, Greenhow and Neufeld (1961) analyzed the horizontal wind above Jodrell Bank (53.2°N, 2.3°W), averaged over the vertical range 80–100 km. They found a diurnal contribution with an ampli-

tude of about 5 m/s, and a semidiurnal contribution with an amplitude of about
13 m/s. Elford (1959) analyzed similar data for the air above Adelaide (34.9 °S,
138.6 °E), and found a diurnal contribution with an amplitude of about 25 m/s; the
amplitude of the semidiurnal contribution was only about 10 m/s. The radio meteor
data thus suggest that the diurnal oscillation may be bigger than the semidiurnal
oscillation, but that it may also be confined to lower latitudes.

Finally, some information about wind oscillations in the upper atmosphere is
obtained from the analysis of daily variations in the geomagnetic field at the ground.
It is thought that these geomagnetic variations (quiet day variations) are caused by
wind-induced electric currents at heights in the neighborhood of 110–120 km. If one
makes various assumptions (often unrealistic) about the spatial structure of the wind,
the temporal and spatial distributions of the electric conductivity, and other matters,
one can estimate the winds causing the geomagnetic variations. This was attempted
by Maeda (1955) and Kato (1956), whose calculated diurnal wind field was twice as
intense as their semidiurnal wind field. However, in contrast to other observations,
both of these inferred wind oscillations increased away from the equator.

1.12. Theoretical Calculations of the Diurnal Thermal Tide

Kato (1966), in order to explain the winds inferred from dynamo calculations, and
Lindzen (1966a), in order to explain the large diurnal oscillations near the stratopause,
closely re-examined the theory of the diurnal thermal tide. Both independently
discovered that for the diurnal oscillations (where $\sigma = \omega$ and $s = 1$), the solutions to
Equation (7) consist in two sets of functions: one associated with negative values of
h, the other associated with small positive values of h. From Equation (11) we see
that negative h implies negative μ^2, and hence, energy trapping, while small positive
h's are generally associated with positive μ^2 and energy propagation. Those modes
associated with negative h have most of their amplitude confined to latitudes poleward
of 30°; the modes with positive h have most of their amplitude confined to latitudes
equatorwards of 30°. These results represent an extension to a spherical atmosphere
of a well known result for internal gravity waves in a plane atmosphere rotating at
rate ω; namely, that internal gravity waves with frequency σ cannot propagate
vertically if $\sigma < 2\omega$; on a rotating sphere $2\omega \cos\theta$ (rather than 2ω) is the relevant
quantity (Eckart, 1960). Lindzen (1967a) calculated the detailed atmospheric
response to diurnal excitation by ozone and water vapor heating. These calculations
accounted for over two thirds of $S_1(p)$ observed at the ground, and also substantially
accounted for many features of the observed diurnal wind oscillations below 100 km.
Lindzen found that most of the thermal excitation goes into the main trapped mode
(mode with negative h). This mode is associated with a significant response in the
region of excitation; however, it does not permit the propagation of a disturbance
away from the region of excitation. This, therefore, explains why most of the diurnal
oscillation excited by ozone heating does not reach the ground. A smaller, but still
significant, part of the diurnal excitation goes into the main propagating mode

(mode with positive h). The vertical wavelength of this mode (approximately 25 km) is short compared to the thickness of the region of ozone heating (approximately 40 km). Thus, the disturbance in this mode excited by ozone heating is, indeed, subject to some destructive interference. However, the same is not true for the disturbance excited by the thinner region of water vapor heating in the troposphere (Green, 1965). In fact, outside of regions of local thermotidal excitation, the diurnal thermotidal oscillation is primarily due to the main propagating mode excited by water vapor heating in the troposphere. This remains true at least until we reach the thermosphere, where viscosity, conductivity and electromagnetic damping may attenuate incoming waves. Details of all these calculations are given in Chapter 3.

1.13. Other Features of Atmospheric Oscillations

It seems that we now have a reliable general theory of S through most of the atmosphere; the cause is almost wholly thermal. Seasonal variations are not yet fully explained, though Butler and Small have shown how that of $S_3(p)$ is probably caused. Also, little theoretical work has been done on the problem of what becomes of thermal tides above about 100 km, where the effects of viscosity, conductivity, hydromagnetic processes and nonlinearity are potentially important.

The annual mean lunar atmospheric tide is about 4 times the equilibrium tide; the gravitational component of $S_2(p)$ is likely to be less than half the small lunar variation $L_2(p)$. The theory for the annual mean of L_2 now appears to be adequate, though it requires that account be taken of atmospheric dissipation (viz. Sections 3.5C and 3.6B). But $L_2(p)$ undergoes a remarkable annual variation, which still needs explanation. Another aspect of $L_2(p)$ not yet adequately studied is the influence of the rise and fall of the oceans on the lunar atmospheric tide. Sawada (1965) discussed this for the case of an ocean that covers the whole earth. He concluded "In spite of this deviation from reality, the result suggests the possibility of a great effect of the ocean on the atmospheric lunar tide. The effects of oceans resembling closely those actually occurring on the earth remain to be discussed" (p. 636).

The part of S_2 that is not a westward travelling wave, following the sun, must depend on spatial inhomogeneities in heating – such as the difference of heating of the atmosphere over the land and seas – but its detailed cause has not been elucidated.

THE SOLAR DAILY ATMOSPHERIC OSCILLATIONS
AS REVEALED BY METEOROLOGICAL DATA

2S.1. The Material Studied; Ground Level Data

The main material for study of the actual atmospheric oscillations consists of series of readings of records of meteorological data – particularly of pressure, but also of wind and temperature – made at ground level at numerous observatories widely distributed over the globe. The readings are commonly taken at hourly intervals, generally at the hours of local mean time, true or standard (that is, of a meridian, at a distance of some degrees of longitude, used for the standard time of the country or region). There is a vast body of such readings; the series for some stations goes back for more than a century. Much of this material is still unused for the purpose here considered.

At most stations the record has long been a continuous one, photographic or by pen recording. But in some places, and especially in earlier times (as at Paris in Laplace's day), they were eye readings, made at longer than hourly intervals. Such series of data, when not at equal intervals throughout the day, need special treatment in the analysis to determine the local manifestation of the atmospheric oscillations; but where better material for a region is lacking, it may be worth while to use them, despite the extra effort involved (see pp. 63, 82).

The first process to be applied to the data is grouping the days, to obtain group means of the series of (for example) hourly values for the days of the group. Published (also unpublished) meteorological tables of hourly values usually give monthly hourly means, and often also annual hourly means. Sometimes the monthly hourly sequences are combined into groups of four, to give mean sequences for the three 'Lloyd' seasons (and the whole year) defined (and here denoted) as follows:

Months included	Name	Symbol
May to August	June solstitial	j
November to February	December solstitial	d
March, April, September, October	Equinoctial	e
All twelve	Yearly	y

Such monthly, seasonal or annual sequences of mean hourly values may be combined for more than one year, giving better-based averages, less affected by the particular vagaries of the individual years.

Such mean daily sequences of values form the material for the study of S. The first step in their use is harmonic analysis, to determine the harmonic components of S, namely S_n; usually only values of n up to 4 are considered.

2S.2. Harmonic Analysis of S; the Non-Cyclic Variation

For various reasons (such as weather and seasonal changes) the variations, for example of barometric pressure p or air temperature T, are not truly periodic from day to day. That is, the value is not the same at the same hour on successive days, or, for example, at the two midnights that begin and end a mean solar day. The change in the course of 24 hours (whatever the initial time) is called the non-cyclic variation (sometimes written, for brevity, ncv).

Let $y(t)$ denote the value at time t of the meteorological element considered, and let y_σ, for $\sigma = 0, 1, 2, ..., r$ denote the value of $y(t_\sigma)$, where

$$t_\sigma = t_0 + \sigma T/r, \tag{1}$$

and T denotes the duration of one day. Here the units in which t and T are expressed will be hours. The non-cyclic variation d for this sequence is given by

$$d = y_r - y_0. \tag{2}$$

When hourly values are used, $r = 24$; sometimes only alternate hourly values (or even every third value) are used, in which case $r = 12$ (or 8).

Usually, following a procedure due to Lamont (1868), the change d is taken to proceed at a uniform rate throughout the day, and the periodic variation S is taken to be represented by $y'(t)$, equal to $y(t)$ with this uniform change removed. Thus $y'(t)$ and the modified sequence y'_σ analyzed are given thus:

$$y'(t) = y(t) - td/T, \qquad y'_\sigma = y'(t_\sigma). \tag{3}$$

Representing $y'(t)$ by a series of harmonic components, thus:

$$y'(t) = A_0 + \sum_n S_n, \qquad S_n = A_n \cos \frac{2\pi n t}{T} + B_n \sin \frac{2\pi n t}{T}, \tag{4}$$

the values of A_0, A_n and B_n are determined as follows. (The difference between (4), and S_n as given on p. 10, is because here t and T are expressed in hour units, whereas there t was expressed in angle at the rate 2π per day.) We first calculate numbers A'_0, A'_n, B'_n that would be the values of A_0, A_n and B_n if t_0 were zero. Thus

$$A'_0 = \frac{1}{r} \sum_{\sigma=0}^{r-1} y_\sigma - \frac{(r-1)d}{2r} \tag{5}$$

$$A'_n = \frac{2}{r} \sum_{\sigma=0}^{r-1} y_\sigma \cos \frac{2\pi n \sigma}{r} + \frac{d}{r} \tag{6}$$

$$B'_n = \frac{2}{r} \sum_{\sigma=0}^{r-1} y_\sigma \sin \frac{2\pi n \sigma}{r} + \frac{d}{r} \cot \frac{n\pi}{r}. \tag{7}$$

Alternatively (Chapman, 1967) the summations may be from $\sigma = 1$ to $\sigma = r$, but then the last term in A'_0 and A'_n must be changed respectively to $-(r+1)\,d/2r$ and $-d/r$. In this case A'_0 and A_0 refer to an epoch T/r later than when (5)–(7) are used.

From these values of A'_n, B'_n we calculate successively s_n, σ_n, σ'_n and A_n, B_n, as follows:

$$s_n^2 = A_n'^2 + B_n'^2, \qquad A'_n = s_n \sin \sigma'_n, \qquad B'_n = s_n \cos \sigma'_n, \tag{8}$$

$$\sigma_n = \sigma'_n + 2\pi n t_0/T, \qquad A_n = s_n \sin \sigma_n, \qquad B_n = s_n \cos \sigma_n. \tag{9}$$

The value of t_0 to be used here depends on the time reckoning to which (4) refers. If it is the local mean time of the station, whose east longitude from Greenwich is ϕ, and if the meridian of time reckoning relative to which the values of y_σ are measured is in east longitude ϕ', and if h is the hour, on this time reckoning, of the value y_0, then

$$t_0 = h + (\theta' - \theta)/15° \tag{10}$$

Sometimes we may wish to use Greenwich or universal time t_u in (4): in that case

$$t_0 = h + \theta'/15°. \tag{10a}$$

When the sequence analyzed is not a sequence of hourly values for one particular day, but a monthly mean sequence, for a month of m days, d will be $(1/m)$ times the (algebraic) increase during the month, or, if many months are combined, the sum of these monthly increases, divided by the total number of days.

If the data are hourly means (cf. p. 80), each A_n, B_n, s_n, must be multiplied by $(n\pi/24)/\sin(n\pi/24)$, namely, for $n = 1, 2, 3, 4$ respectively, 1.003, 1.012, 1.026, 1.047.

2S.3. The Seasonal Variation of S

In general, if monthly mean daily sequences are analyzed for each calendar month of a single year, or from groups of the same calendar month averaged over several years, the values of A_0, A_n, B_n, s_n, σ_n will differ from one calendar month to another. The differences represent a seasonal (plus some accidental) variation. The change from month to month throughout the year may be shown by a polygon of monthly or seasonal points on a harmonic dial, for any S_n, as in Figure 2S.1 for $S_2(p)$.

The seasonal variation can be expressed analytically in a standard way applicable to each S_n, and to the S_n's for different stations, by harmonic analysis of the sequences A_n, B_n for the twelve calendar months, in order. Though the calendar months differ slightly in length, it has hitherto been considered convenient and adequate (having regard to the accidental part of the changes) to ignore the differences in making the harmonic analysis, thus treating the intervals between the mean epochs in successive months as if they were all one-twelfth of a year. Greater accuracy can be gained if Bartels' seasonal day numbers (Section 2S.3A) are used.

Let y_σ, where $\sigma = 1, 2, \ldots, 12$, denote the values of any chosen one of the harmonic constants of S, namely A_0 or A_n or B_n, for the respective months January to December. If we have such a sequence for each of several years, we can determine the non-cyclic

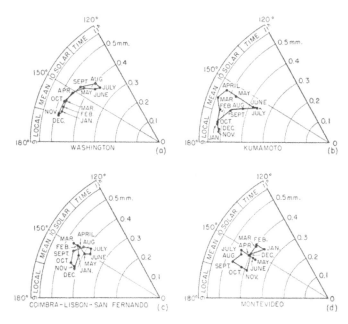

Fig. 2S.1.   Harmonic dials showing the amplitude and phase of $S_2(p)$ for each calendar month for four widely spaced stations in middle latitudes, (a) Washington, D.C., (b) Kumamoto, (c) mean of Coimbra, Lisbon and San Fernando, (d) Montevideo (Uruguay). After Chapman (1951).

variation D for each year, or for the average sequence for all the years (this requires analysis of one extra month at the end). Then harmonic constants α_0, α_m, β_m, expressing the seasonal variation of the particular S_n harmonic constant considered, can be thus calculated:

$$\alpha'_0 = \frac{1}{12} \sum_1^{12} y_\sigma - \frac{13}{24} D \tag{1}$$

$$\alpha'_m = \frac{1}{6} \sum_1^{12} y_\sigma \cos \frac{\pi m \sigma}{6} - \frac{D}{12} \tag{2}$$

$$\beta'_m = \frac{1}{6} \sum_1^{12} y_\sigma \sin \frac{\pi m \sigma}{6} + \frac{D}{12} \cot \frac{m\pi}{12}. \tag{3}$$

Results of such calculations (not considering D) were first given by Siebert (1956a). For $S_2(T)$ for the mean of 8 European stations between latitudes $50°$ and $60°N$ he obtained the polygon of monthly dial points shown in Figure 2S.2(a). From the 12 values of A_2 and B_2 he calculated harmonic constants α_0, α_m, β_m for A_2 and for B_2: he found the values for $m = 3$ or more to be negligible. His values of α_0 for A_2 and B_2 represent the annual mean $S_2(T)$, shown by the vector in Figure 2S.2(b). The additional calendar monthly vectors given by $\alpha_m \cos(\pi m\ \sigma/6) + \beta_\gamma \sin(\pi m\ \sigma/6)$ for A_2 and for B_2 lie on an ellipse: Figure 2S.2(b) shows these ellipses for $m = 1$ (the annual component variation of S_2) and $m = 2$ (the semi-annual component variation);

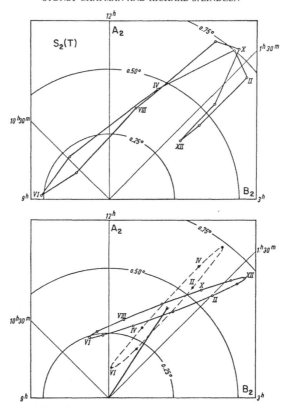

Fig. 2S.2.   Harmonic dials showing (above) the monthly mean dial vectors of $S_2(T)$ for the mean of the latitude belt 50° to 60°N; and (below) the resolution of the month to month variation of $S_2(T)$ into a variation of period 1 year (the full-line ellipse) and one of period 1/2 year (the broken-line ellipse, described twice each year). After Siebert (1956a).

the latter ellipse is described twice in each year, and the points on it marked I, II, ... (for January, February, ...) correspond also to VII, VIII, ... (for July, August, ...).

Such harmonic studies of the seasonal variation of the components S_n of S for meteorological and other geophysical elements have barely been begun. They should be undertaken for $S_2(p)$ for comparison with the dynamical theory of its seasonal variation, which likewise has yet been little studied.

2S.3A. DAILY SEASONAL INTEGERS σ (SIGMA) OR SN (BARTELS, 1954)

The varying incidence of sunlight upon the earth in the course of a year, depending on the declination of the sun, powerfully affects meteorological, geomagnetic, and ionospheric phenomena. The 12-monthly division of the year in the present Gregorian calendar is not well adapted to the study of this varying solar influence; the months and quarters are unequal in length, and the solstices and equinoxes neither delimit the months, nor are central to them. Hence a new and more logical division of the year is desirable for use in geophysical investigations. For this purpose a daily seasonal integer Sigma (σ) may be assigned to each calendar day (including February 29 in Leap Years), the same for every year – a perpetual Sigma calendar, which can remain unaffected by any future general calendar reform, with the appropriate redefinition of the Sigmas.

The Sigma integers σ are 24 odd numbers ranging from 1 to 47 (the numbers 47, 45, ... might also be considered as equivalent to $-1, -3, ...$). Each σ applies to all the days of a group of 15 or 16; the first date and number of days for each group are as follows:

σ for group	Initial date	Days in group	σ for group	Initial date	Days in group	σ for group	Initial date	Days in group
01	Dec. 22	15	17	Apr. 22	15	33	Aug. 21	16
03	Jan. 6	15	19	May 7	15	35	Sep. 6	15
05	Jan. 21	15	21	May 22	15	37	Sep. 21	15
07	Feb. 5	15	23	June 6	15	39	Oct. 6	16
09	Feb. 20	15 or 16	25	June 21	15	41	Oct. 22	16
11	Mar. 7	16	27	July 6	15	43	Nov. 7	15
13	Mar. 23	15	29	July 21	16	45	Nov. 22	15
15	Apr. 7	15	31	Aug. 6	15	47	Dec. 7	15

These 24 groups are based on the longitude h of the mean sun; σ represents the odd integer nearest to $48(h° - 90°)/360°$. The group 1 is chosen to begin at the winter solstice, so that the series 1 to 47 should closely approximate to a calendar year. The number of days in a calendar year, 365 or 366, precludes a strictly regular progression of σ throughout the year, but the progression chosen is as even as possible.

The 24 groups can be combined into 12 pairs, beginning with the pair 47, 1 (or -1, 1), forming a set of 30 days centered on December 21 midnight; this set may be numbered 0, and the later sets 4, 8, 12, ..., 44. This division of the year is analogous to, but better than, the customary division of the calendar year into the 12 Gregorian months.

The 24 groups can also be combined into 8 sets of 3 groups each, namely sets 1, 3, 5; 7, 9, 11; .... This divides the year into octants, which may be numbered 3, 9, 15, ..., 45. This division of the year may be appropriate when the data available do not justify the 12-fold division of the year, but more than justify a grosser sub-division, into only 6 or 4 sets of groups. The division into octants suffices to enable annual and semiannual variations to be determined, in any geophysical property evaluated for each octant, such as daily means, daily ranges, or Fourier coefficients of any component of a daily variation, whether solar, lunar or lunisolar.

Grosser combinations of the 24 groups are possible into 6 sets of 4 groups each: or 4 sets of 6 groups: or 3 sets of 8 groups: or 2 sets of 12 groups.

If it is preferred to have 8 sets centered at the solstices, the equinoxes, and midway between those epochs, this may be done by taking the group with center 0 to consist of the groups -1 and 1, together with the groups -3 and 3 taken with half weight, the combination being divided by three; similarly, the group with center 6 would consist of $(+3)/2+(+5)+(+9)/2$ divided by 3; and so on. This use of the intervening groups 3, 9, 15,... twice over, with half weight each time, produces a slight smoothing; this may be welcome where relatively few days are available. These 8 sets of groups can be combined, if desired, into 4 sets of groups, likewise centered at 0, 12, 24, and 36.

The alternate threefold division of the year, introduced by Lloyd and much used in the past in geomagnetism, is into three four-monthly groups, of which two are centered near the solstices, and the third is made up of two months near each of the equinoxes; thus the third group does not consist of four consecutive months. These groups have been commonly denoted by j (May-August), the June solstitial group, d (November-February), the December solstitial group, and e (March, April, September, October), the equinoctial group. When Sigma numbers are used, the appropriate

TABLE 2S.1

Jan.	Feb.	Mar.	Apr.	May	Jun.	Jul.	Aug.	Sep.	Oct.	Nov.	Dec.
1 01	1 05	1 09	1 13	1 17	1 21	1 25	1 29	1 33	1 37	1 41	1 45
2 01	2 05	2 09	2 13	2 17	2 21	2 25	2 29	2 33	2 37	2 41	2 45
3 01	3 05	3 09	3 13	3 17	3 21	3 25	3 29	3 33	3 37	3 41	3 45
4 01	4 05	4 09	4 13	4 17	4 21	4 25	4 29	4 33	4 37	4 41	4 45
5 01	5 07	5 09	5 13	5 17	5 21	5 25	5 29	5 33	5 37	5 41	5 45
6 03	6 07	6 09	6 13	6 17	6 23	6 27	6 31	6 35	6 39	6 41	6 45
7 03	7 07	7 11	7 15	7 19	7 23	7 27	7 31	7 35	7 39	7 43	7 47
8 03	8 07	8 11	8 15	8 19	8 23	8 27	8 31	8 35	8 39	8 43	8 47
9 03	9 07	9 11	9 15	9 19	9 23	9 27	9 31	9 35	9 39	9 43	9 47
10 03	10 07	10 11	10 15	10 19	10 23	10 27	10 31	10 35	10 39	10 43	10 47
11 03	11 07	11 11	11 15	11 19	11 23	11 27	11 31	11 35	11 39	11 43	11 47
12 03	12 07	12 11	12 15	12 19	12 23	12 27	12 31	12 35	12 39	12 43	12 47
13 03	13 07	13 11	13 15	13 19	13 23	13 27	13 31	13 35	13 39	13 43	13 47
14 03	14 07	14 11	14 15	14 19	14 23	14 27	14 31	14 35	14 39	14 43	14 47
15 03	15 07	15 11	15 15	15 19	15 23	15 27	15 31	15 35	15 39	15 43	15 47
16 03	16 07	16 11	16 15	16 19	16 23	16 27	16 31	16 35	16 39	16 43	16 47
17 03	17 07	17 11	17 15	17 19	17 23	17 27	17 31	17 35	17 39	17 43	17 47
18 03	18 07	18 11	18 15	18 19	18 23	18 27	18 31	18 35	18 39	18 43	18 47
19 03	19 07	19 11	19 15	19 19	19 23	19 27	19 31	19 35	19 39	19 43	19 47
20 03	20 09	20 11	20 15	20 19	20 23	20 27	20 31	20 35	20 39	20 43	20 47
21 05	21 09	21 11	21 15	21 19	21 25	21 29	21 33	21 37	21 39	21 43	21 47
22 05	22 09	22 11	22 17	22 21	22 25	22 29	22 33	22 37	22 41	22 45	22 01
23 05	23 09	23 13	23 17	23 21	23 25	23 29	23 33	23 37	23 41	23 45	23 01
24 05	24 09	24 13	24 17	24 21	24 25	24 29	24 33	24 37	24 41	24 45	24 01
25 05	25 09	25 13	25 17	25 21	25 25	25 29	25 33	25 37	25 41	25 45	25 01
26 05	26 09	26 13	26 17	26 21	26 25	26 29	26 33	26 37	26 41	26 45	26 01
27 05	27 09	27 13	27 17	27 21	27 25	27 29	27 33	27 37	27 41	27 45	27 01
28 05	28 09	28 13	28 17	28 21	28 25	28 29	28 33	28 37	28 41	28 45	28 01
29 05		29 13	29 17	29 21	29 25	29 29	29 33	29 37	29 41	29 45	29 01
30 05		30 13	30 17	30 21	30 25	30 29	30 33	30 37	30 41	30 45	30 01
31 05		31 13		31 21		31 29	31 33		31 41		31 01

grouping is as follows:

d: $(41) + (43) + (45) + (47) + (1) + (3) + (5) + (7)$

j: $(17) + (19) + \ldots (31)$

e: $(9) + (11) + (13) + (15) + (33) + (35) + (37) + (39)$.

The entry of the Sigma integers on to punched cards requires 2 columns, as given in Table 2S.1. In a Leap Year all the numbers are as usual, and February 29 is an additional day in the σ group 09. One column on the cards, however, will suffice if the division of the year is not to exceed 12 sets; each set can then be characterized by the quotient of its initial Sigma integer by the number of groups in each set; e.g., for a 12-fold division the sets would be numbered 0, 1,..., 9, 10(T), 11(E).

2S.4. The World-Wide Distribution of S, Particularly of $S(p)$

The world-wide distribution of any harmonic component of a daily variation, e.g., that of $S_2(p)$, can be represented graphically or analytically, in different ways. For example, Haurwitz (1956) has given Figures 2S.3a, b showing on one map equilines of s_2, and on the other, equilines of σ_2; alternatively maps could be drawn showing equilines of A_2 and B_2. From the latter, if the values are read at a regular network of points, where uniformly spaced meridians and latitude circles intersect, it is possible to use them to express the distribution of $S_2(p)$ in terms of spherical harmonic functions. This was first done by Schuster (1889) for the components S_n of the solar daily geomagnetic variation; for the method cf. *Geomagnetism* (Chapman and Bartels, 1940, Chapters 17, 20), or Siebert (1961, Section 2.2). The details are not given here.

2S.4A. $S_2(p)$

The component daily meteorological variation that has aroused most interest is $S_2(p)$, which has been studied by Angot (1887), Hann (1889–1918b), Schmidt (1890, 1921), Simpson (1918) and others. Simpson used data from 214 stations for the annual mean $S_2(p)$. Following a suggestion by Schmidt (1890), he expressed it as follows, as the sum of a wave that travels daily round the earth (a sine function of local time t with coefficient and phase independent of the longitude ϕ) and a zonal oscillation that is a sine function of universal time t_u ($=t-\phi$), with coefficient and phase independent of longitude:

$$S_2(p) = s_2 \sin(2t + \sigma_2) = b \sin(2t + \beta) + c \sin(2t_u + \gamma). \tag{1}$$

Simpson determined b, β, c, γ from $S_2(p)$ data for 190 stations for each of 8 sets divided according to latitude, as in Table 2S.2.

The phase β of the travelling wave has a remarkably small range ($9°$ or ± 9 min of time of maximum) in the eight zones. The amplitude b, decreasing steadily polewards

TABLE 2S.2

Constants for the representation of $S_2(p)$ at various latitudes (Simpson, 1918)

Group		Travelling wave		Zonal wave	
Mean lat.	No. of stations	$b(10^{-2}$ mb$)$	$\beta°$	$c(10^{-2}$ mb$)$	$\gamma°$
0°	17	122.7	156.8	8.1	−4.0
18	15	111.3	155.3	10.9	−23.2
30	12	82.7	149.1	7.9	10.4
40	46	51.6	153.9	5.7	91.1
50	60	30.7	153.0	5.5	104.4
60	18	12.8	158.0	8.3	108.4
70	14	} 2.9	152.9	9.6	98.6
80	8			10.7	116.4
		Mean	154.1		

from the equator, is well represented by the formula

$$b = 0.937 \sin^3 \theta \text{ mm} = 1.25 \sin^3 \theta \text{ mb}, \tag{2}$$

where θ denotes the colatitude.

Up to about 50° latitude the zonal wave, here denoted by $S_2^0(p)$, makes only a minor contribution to $S_2(p)$, but in high latitudes it is predominant. Wilkes (1949) represented it approximately thus (in mb units):

$$0.093 \sin 2t_u + |\cos \theta| (-0.13 \sin 2t_u + 0.10 \cos t_u). \tag{3}$$

Here $|\cos \theta|$ denotes the positive magnitude of $\cos \theta$. Haurwitz (1956, p. 25) has remarked that the first term here implies that the average p for the whole earth changes periodically, "which seems unlikely".

A much more extensive study of $S_2(p)$ was made by Haurwitz (1956). Using data for 296 stations, he gave the (slightly smoothed) maps of equilines of the annual mean amplitude s_2 and phase σ_2 (relative to local mean solar time) shown in Figures 2S.3a, b. The singular points of convergence of the phase equilines arise from the combination of the main travelling wave and the standing or zonal wave, and had previously appeared in a corresponding map given by Simpson. Besides these two main component waves, however, the $S_2(p)$ distribution includes many other waves; all but one of these is of decidedly smaller amplitude.

Haurwitz discussed the local irregularities of $S_2(p)$ as shown by Figure 2S.3, noting the region of small phase angle near the Gulf of Guinea, and also that over the western United States, which had previously been indicated by Spar (1952) in a study of

Fig. 2S.3.   World maps showing equilines of (below) the amplitude (s_2, unit 10^{-2} mb) and (above) the phase (σ_2) of $S_2(p)$, relative to local mean time. After Haurwitz (1956).

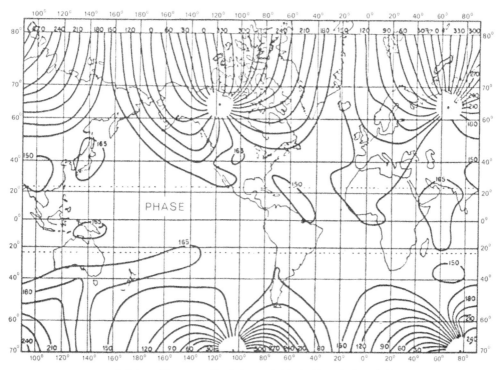

$S_2(p)$ equilines: amplitude (s_2; unit 10^{-2} mb) and phase (σ_2)

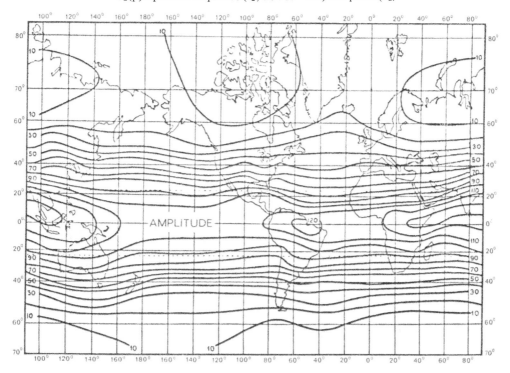

$S_2(p)$ at 100 stations over north America; see also Figure 1.4. The latter anomaly is attributed to the Rocky Mountains. Spar concluded that the topography affects the phase more than the amplitude, and that the influence of the Cordilleras resembles that inferred theoretically by Kertz (1951). In high latitudes the phases are mainly controlled by the zonal wave; Haurwitz and Sepúlveda (1957) made a special study of $S_2(p)$ and its seasonal variation in the Arctic.

Using readings from Figure 2S.3 at a regular network of 384 points, 15° apart in longitude, round latitude circles at 10° intervals from 80°N to 70°S, Haurwitz (1956) used the method of least squares to obtain, for each latitude, the four constants b, β, c, γ, in (1) that best fit these numbers, after converting them to the A_2, B_2 coefficients, functions of θ and ϕ. He gave Table 2S.3, similar to Table 2S.2, but for 16 latitudes

TABLE 2S.3

Amplitude and phase angle of W_2 and Z_2 (Haurwitz, 1956)

Lat.	W_2		Z_2	
	b	β	c	γ
	10^{-2} mb		10^{-2} mb	
80°N	1.0	148.3°	10.2	109.8°
70°	3.4	146.3°	9.1	107.9°
60°	12.5	154.3°	8.0	102.4°
50°	28.4	153.4°	7.0	92.0°
40°	52.1	153.7°	2.7	78.8°
30°	73.7	153.2°	2.0	304.3°
20°	95.8	156.6°	6.9	331.9°
10°N	113.7	157.4°	5.7	350.1°
0°	117.5	158.2°	3.8	24.8°
10°S	110.5	159.5°	4.2	73.0°
20°	96.4	160.7°	2.8	26.9°
30°	77.8	159.6°	1.7	105.9°
40°	50.1	160.8°	5.7	127.1°
50°	25.2	159.9°	7.7	102.3°
60°	9.2	151.6°	5.0	120.7°
70°S	0.9	162.8°	5.0	120.8°

10° apart, and found that the formula for the travelling wave

$$1.16 \sin^3 \theta \sin(2t + 158°) \text{ mb},$$

fits his b, β data very well; both the amplitude and phase agree rather well with those given by Simpson. Haurwitz also gave a more complicated formula for this wave, involving three spherical harmonic functions; but the fit with the data was not more than equally good.

Haurwitz expressed the zonal part of $S_2(p)$ in the form (analogous to Simpson's) (in mb)

$$0.085 \, P_2(\theta) \sin(2t_u + 118°), \quad P_2 = (3\cos^2\theta - 1)/2.$$

The amplitude here is only two thirds of that given by Simpson, Haurwitz also expressed this zonal wave as a sum of four terms of the form $c_k E_k(\theta) \sin(2t_u + \sigma_k)$, for k from 1 to 4; the Hough functions E_k are of the special ($s \doteq 0$) Solberg (1936) type

$$E_{2k+1} = \sin\left(\frac{2n-1}{2}\cos\theta\right) \qquad E_{2k} = \cos(k\cos\theta),$$

indicated by Haurwitz (1956) in his theoretical study of $S_2(p)$ and its relation to $S_2(T)$.

Kertz (1956b) determined additional waves in $S_2(p)$, using the tabulated values of A_2 and B_2 derived by Haurwitz from Figure 2S.3. Kertz harmonically analyzed each series of 24 values round each circle of latitude, giving:

$$A_2(\theta, \phi) = \sum_\nu (k_\nu \cos\nu\phi + l_\nu \sin\nu\phi) \qquad (4a)$$

$$B_2(\theta, \phi) = \sum_\nu (m_\nu \cos\nu\phi + n_\nu \sin\nu\phi). \qquad (4b)$$

Then he converted the expression for $S_2(p)$ relative to local mean solar time to an expression relative to Greenwich or universal time t_u, made up of waves W_2^s each of the form

$$\alpha_s \cos(2t_u + s\phi) + \beta_s \sin(2t_u + s\phi), \quad \text{or} \quad \gamma_s \sin(2t_u + s\phi + \varepsilon_s). \qquad (5)$$

The factors α_s and β_s are given by the following scheme:

$$
\begin{array}{llll}
s: & \text{less than 2} & 2 & \text{greater than 2} \\
\alpha_s: & (k_{2-s} + n_{2-s})/2 & k_0 & (l_{s-2} - m_{s-2})/2 \\
\beta_s: & (-k_{2-s} + n_{2-s})/2 & m_0 & (l_{s-2} + m_{s-2})/2.
\end{array}
$$

In the Equations (4), (5) the coefficients k, l, m, n and α_s, β_s and the amplitude γ_s are all functions of θ. Mean amplitudes $\bar\gamma_s$ were calculated as follows, from the 16 values of the γ_s's (cf. Kertz's Equation (2.5)):

$$\bar\gamma_s^2 = \sum \gamma_s^2 / 16. \qquad (6)$$

Figure 2S.4 shows on a logarithmic scale the values of $\bar\gamma_s$ for the larger amplitudes. The travelling wave corresponds to $s=2$, and is outstanding; its value agrees well with that deduced by Haurwitz from his maps. The zonal wave is the one for which $s=0$, and this is the next greatest; but the wave for which $s=3$ has nearly as great an amplitude. All the other terms are much smaller. The magnitude of the wave W_2^3 was first revealed by this study by Kertz.

Each set of coefficients $\alpha_s(\theta)$ and $\beta_s(\theta)$, for the 16 values of θ considered, can be expressed as a sum of associated Legendre functions, of order 2 and degree 2 or more.

2S.4A.1. *Types of associated Legendre Functions*

Different types of associated Legendre functions, differing only by numerical factors, are in use in mathematical and geophysical literature, and different notations are in use for the same or different forms. Schmidt (1935; see also Chapman and Bartels, 1940, pp. 611, 612) introduced a form, here denoted by $P_n^m(\theta)$, which he called semi-normalized; it is much in use in geomagnetic studies, and has also been used in studies of atmospheric tides and thermal tides by Haurwitz (1956, 1965), Kertz (1956a) and Siebert (1961). In a study of $L_2(p)$, Haurwitz and Cowley (1970) have used a normalized type, here denoted by $P_{n,m}(x)$, where

$$x = \cos \theta,$$

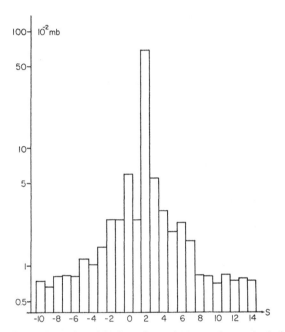

Fig. 2S.4.   The amplitudes (on a logarithmic scale, and averaged over the latitudes 80° N to 70° S) of the semidiurnal pressure waves, parts of $S_2(p)$, of the type $\gamma_s \sin(2t_u + s\phi + \sigma_s)$, where t_u signifies universal mean solar time. After Kertz (1956b).

and the same form is used later in this review. The two forms are thus related: for

$$m = 0, \quad P_{n,m}/P_n^m = (n + \tfrac{1}{2})^{1/2}; \quad \text{for } m > 0, \quad P_{n,m}/P_n^m = (2n + 1)^{1/2}/2.$$

Tables of values of P_n^m and associated quantities have been given by Schmidt (1935), and Belousov (1962) has tabulated $P_{n,m}$.

The type $P_{n,m}$ may be called normalized because

$$\int_{-1}^{1} \{P_{n,m}(x)\}^2 \, dx = 1.$$

The functions $S_n^m(\theta, \phi)$ defined by $P_n^m \cos m\phi$ and $P_{n,m} \cos m\phi \ (= S_{n,m})$ are spherical harmonic surface functions, and their root mean square values over a sphere of unit radius (hence of area 4π), is $1/(2n+1)^{1/2}$ for S_n^m, and $\tfrac{1}{2}$ for $S_{n,m}$. The same is true, of course, if $\cos m\phi$ here is replaced by $\sin m\phi$. Thus the root mean square value of every function $S_{n,m}$ over the unit sphere is the same.

In discussing the expressions of the atmospheric tides and thermal tides in terms of associated Legendre functions, in this chapter the Schmidt type is used. In Chapters 2L and 3 both forms N_n^m and $P_{n,m}$ are used.

The functions P_n^m are related as follows to the Neumann form, here denoted by N_n^m and thus defined, for $m>0$ (for $m=0$, $P_n^m = N_n^m = P_n$).

$$N_n^m = (\sin\theta)^m \frac{d^m P_n}{d(\cos\theta)^m}, \quad P_n^m = \left\{2\frac{(n-m)!}{(n+m)!}\right\}^{1/2} N_n^m,$$

where P_n $(=P_n^0)$ signifies the Legendre function. Explicit formulae for P_n^m are given by Matsushita and Campbell (1967, pp. 1353–1355) for m from 0 to 6, and for nine values of $n(\geqslant m)$ for each m; see also Chapman and Bartels (1940; chap. 17).

2S.4A.2. The Spherical Harmonic Expression of $S_2(p)$

Kertz (1956b) expressed the three main waves in $S_2(p)$, namely W_2^2, W_2^0, W_2^3, in spherical harmonic terms, using Schmidt's form of the associated Legendre functions. This involves the expression of the series of α_s and β_s for the 16 values of θ as series of such functions. Kertz in an appendix explained his method of deriving the coefficients and phases of the several terms of the form $c_k^s P_k^s(\theta) \sin(2t_u + s\phi + \varepsilon_k^s)$, and how to estimate the probable errors (though he did not give such estimates). Table 2S.4 gives the terms he found, that have amplitudes c at least equal to 0.04 mb. The columns marked (H) give the corresponding values deduced by Kertz from Table 2S.3 (of Haurwitz). The agreement is good.

Figure 2S.5 shows the isobars for the main harmonic term $(k=3)$ of the wave $W_{2,3}^3$

TABLE 2S.4, $W^s{}_{2,k}$

		$100\,c$ (mb)	ε	$100c(H)$ (mb)	$\varepsilon(H)$
$s=2$	$k=2$	122	158°	123	158°
	$k=3$	4.6	78	4.5	77
	$k=4$	21.6	342	22.5	343
	$k=6$	4.3	332		
$s=0$	$k=2$	7.2	135	8.5	118
	$k=3$	5.6	123		
	$k=6$	4.9	275		
$s=3$	$k=3$	10.7	88		
	$k=7$	4.1	183		

at Greenwich time 0 h or noon. This system of three high and three low pressure areas moves round the earth in 1.5 days; its dependence on θ, according to $\sin^3\theta$, is the same as that of the main travelling wave W_2^2.

Kertz also expressed the main terms W_2^3, W_2^0 and W_2^2 in Hough functions, both

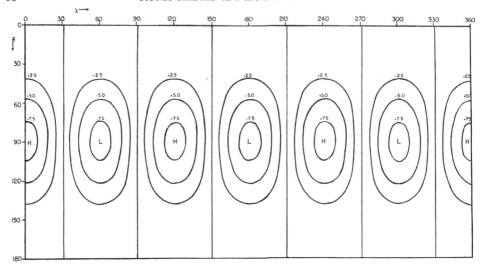

Fig. 2S.5.   Isobars of the wave $W^3{}_{2,3}(p)$, proportional to $P_3{}^3 \sin(2t_u + 3\phi + \sigma_3)$, the component of $S_2(p)$ next in importance after the travelling and standing waves $s = 2$, $s = 0$. Unit 10 μb.
After Kertz (1956a).

for p and T, and in connection therewith he discussed atmospheric resonance conditions.

The $S_2(p)$ data so far considered are for the annual mean; parallel studies for the seasons j, e, d have not so far been made. The variation of $S_2(p)$ throughout the year is shown by the sets of monthly dial points in Figure 2S.1, for 5 northern stations and one southern (nos. 24, 37, 21 + 25 + 31, 99 of Table 2L.2). The maximum occurs earlier in winter (d) than summer (j) at the northern stations.

2S.4B. $S_1(p)$

The first comprehensive study of $S_1(p)$ as an aspect of a worldwide atmospheric oscillation was made by Haurwitz (1965). He had data for 228 stations, very irregularly distributed over the globe – with large gaps over the Pacific Ocean, and only a few stations for the southern latitudes from 40° to 90°. Most of the harmonic constants available to him were based on five or more years of record, but for a few high latitude stations the record covered only about one year. His sources included Hann's extensive compilations (mainly in 1889, but supplemented by Hann in 1892, 1917 and 1918), and others by Schou (1939) for northern Europe, by Frost (1960) for Malaya, by Sellick (1948) for southern Africa, and by weathership data collected by Rosenthal and Baum (1956) for the North Atlantic. Where harmonic coefficients for any station were available from more than one analysis, those based on the

Fig. 2S.6.   World maps showing equilines of the coefficients A_1 (below) and B_1 (above) of $S_1(p)$, relative to local mean time, unit 10^{-2} mm Hg. After Haurwitz (1965), wherein the captions of Figures 1, 2 should be interchanged.

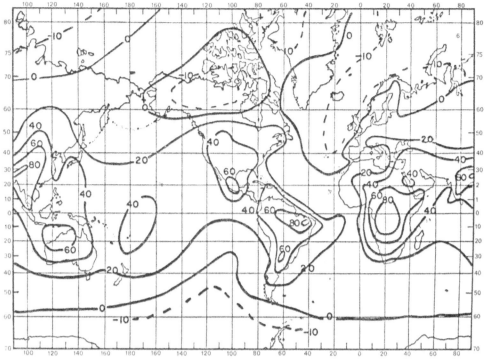

S_1 (p) equilines, A_1 (below) and B_1 (above; u is 10^{-2} mb).

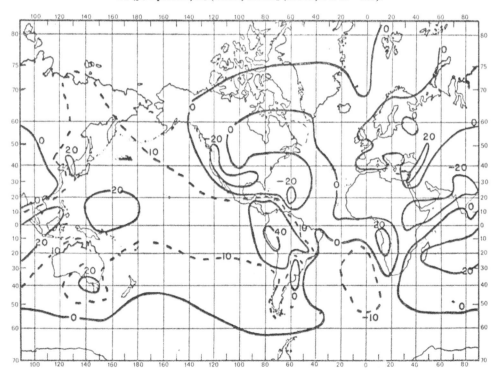

longer record were always chosen. In general the probable errors of the determinations of the first harmonic of $S(p)$ were not available.

The geographical distribution of $S_1(p)$ was represented on world maps by equilines – one map for each harmonic coefficient, A_1 and B_1. These are reproduced here as Figures 2S.6a, b, which show equilines at intervals of 2 mm Hg (the maps actually drawn for the later analysis had equilines at 1 mm intervals). In drawing the equilines, cases where special local conditions at a station made $S(p)$ there very abnormal were disregarded.

The maps clearly show the influence of the continents on $S_1(p)$. In general, A_1 exceeds B_1, and equilines of the amplitude s_1 would not be significantly less irregular than those for A_1. Hence comparison of Figure 2S.6a with Figure 2S.3a for s_2 well shows how much more regular is the distribution of $S_2(p)$ than that of $S_1(p)$. Nevertheless, by and large, s_1 does decrease with increasing latitude.

For the purpose of spherical harmonic analysis, Haurwitz used A_1 and B_1 values read from the maps at 336 ($= 24 \times 14$) grid points, along 14 circles of latitude at $10°$ intervals, from $10°$ to $140°$ colatitude, and at intervals of $15°$ of longitude along each latitude circle. For each circle a harmonic analysis was made of the 24 values of A_1 and of B_1, along the circle. Thus harmonic series for each colatitude θ_r were obtained as follows:

$$A_1 = \sum_v (k_v \cos v\phi + l_v \sin v\phi) \tag{1}$$

$$B_1 = \sum_v (m_v \cos v\phi + n_v \sin v\phi). \tag{2}$$

From these, following Kertz (1956b, 1959), component waves were obtained, expressed thus in terms of UT:

$$\alpha_s \cos(t_u + s\phi) + \beta_s \sin(t_u + s\phi). \tag{3}$$

For $s = 1$, $\alpha_s = k_0$, $\beta_s = m_0$; for $s < 1$, $\alpha_s = (k_{1-s} + n_{1-s})/2$, $\beta_s = (m_{1-s} - l_{1-s})/2$, and for $s > 1$, $\alpha_s = (k_{s-1} - n_{s-1})/2$, $\beta_s = (m_{1-s} + l_{1-s})/2$.

The amplitudes and phases of the terms with wave numbers (s) from -7 to $+9$ were determined. The amplitude γ_s (where $\gamma_s^2 = \alpha_s^2 + \beta_s^2$) was found for each term for each colatitude θ_r from $10°$ to $140°$: and from these 14 values of γ_s, a mean amplitude $\bar{\gamma}_s$ was calculated for each s, using the formula

$$\bar{\gamma}_s = \{\sum_r \gamma_s(\theta_r) \sin \theta_r\}/\{\sum_r \sin \theta_r\}. \tag{4}$$

Here the factor $\sin \theta_r$ allows for the different areas of the $10°$ latitude belts. This method of determining $\bar{\gamma}_s$ differs from that used by Kertz for $S_2(p)$; cf. (6) of Section 2S.4A. Figure 2S.7 shows $\bar{\gamma}_s$ for each s, the unit being the microbar (μb).

By far the largest amplitude is that of the term for $s = 1$, corresponding to a wave travelling westward with the sun, proportional to the local time function $\sin(t + \sigma_1)$. All the other waves have much smaller amplitude, including the zonal wave $s = 0$, dependent only on UT and latitude.

The geographical distribution of the coefficients α_s, β_s for each combination $\pm s$

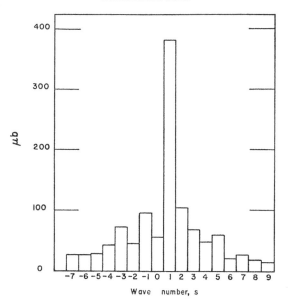

Fig. 2S.7.   The amplitudes (on a logarithmic scale, and averaged over the latitudes from the north pole to 60°S) of the diurnal pressure waves, parts of $S_1(p)$, of the type $\gamma_s \sin(t_u + s\phi + \sigma_s)$. After Haurwitz (1965).

was then represented approximately by only three spherical harmonic terms, of order $|s|$ and degree $|s| + i$, of the Schmidt type, for $i = 0, 1, 2$, namely

$$\gamma^{|s|}_{|s|+i} \, P^{|s|}_{|s|+i} \sin (t_u + s\phi + \sigma^{|s|}_{|s|+i}).$$

The coefficients of these several harmonic terms were determined by the method of least squares, with weights $\sin\theta_r$ for the circle of colatitude θ_r, from the series of values of α_s and β_s. Then their amplitudes and phases were found. Those for the four waves specially considered, namely those for $s = -1, 0, 1, 2$, are given in Table 2S.5.

The representation of $S_1(p)$ by only these twelve spherical harmonic terms ignores many waves of small mean amplitude, such as are indicated in Figure 2S.7. Thus the 12-term series can give only a smoothed picture of the geographical distribution of $S_1(p)$.

The sum of the 8 principal wave components in Table 2S.5, ignoring the four terms whose amplitudes are less than 40 μb, gives values A_1^* and B_1^* for A_1 and B_1 for each of the 336 grid points on the map. Their root-mean-square difference from the values read from the maps, namely

$$\left[\frac{\sum \{(A_1^* - A_1)^2 + (B_1^* - B_1)^2\}}{336} \right]^{1/2} \tag{5}$$

is 209 μb (or, for comparison with Figure 2S.6b, 15.6×10^{-2} mm Hg).

Even only the two principal terms in Table 2S.5, with amplitudes 464 and 210 μb, which relate to the main westward travelling wave S_1^1, give a fairly good approxima-

TABLE 2S.5

(Haurwitz, 1965)

Wave Number	$|s| + i$	Amplitude (μb) $\gamma_{|s|+i}^{|s|}$	Phase $\sigma_{|s|+i}^{|s|}$
$s = 1$	1	464	12°
	2	69	321
	3	210	195
$s = 2$	2	158	263
	3	69	235
	4	4	293
$s = -1$	1	19	279
	2	134	109
	3	40	174
$s = 0$	0	12	154
	1	16	301
	2	61	182

tion to the actual geographical distribution of $S_1(p)$; the root-mean-square difference using this simplified representation is 248 μb, or 18.5×10^{-2} mm Hg.

As these two principal terms are almost opposite in phase (the difference being $195° - 12° = 183°$), they can with little error be combined, and their sum is well approximated by the very simple formula

$$S_1^1(p) = 593 \ \mu b \ \sin^3 \theta \ \sin(t + 12°), \tag{6}$$

expressed in terms of local time t.

Figure 2S.8 shows how well (6) represents the mean amplitude s_1 for each latitude,

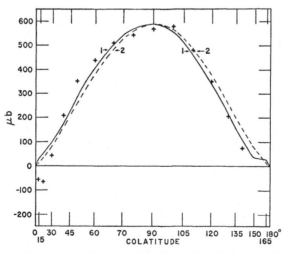

Fig. 2S.8.   The crosses show the amplitude of $S_1^1(p)$ for each latitude; curve 1 (full line) shows the amplitude given by the sum of 3 terms (S_1^1, S_2^1, S_3^1), and curve 2 (broken line) shows the amplitude expressed as a multiple of $\sin^3\theta$, where θ denotes the colatitude. After Haurwitz (1965).

derived from the first *three* $S_1^1(p)$ spherical harmonics P_m^s (9 terms) in Table 2S.5. The crosses show the mean amplitude $\bar{s}_1(\theta_r)$ of $S_1(p)$ for each latitude, derived from the maps (Figures 2S.6a, b). The formula (6) approximately represents the main westward travelling wave S_1^1 of $S_1(p)$; it shows that this has an amplitude that varies with latitude in the same way as does the main travelling wave S_2^2 of $S_2(p)$, according to Simpson's formula (2) of Section 2S.4A. According to these formulae, the amplitude of S_1^1 at the equator (593 μb) is rather less than half that of S_2^2 at the equator (1250 μb).

The corresponding main travelling wave S_1^1 of $S(T)$, the solar diurnal variation of air temperature at ground level, has an amplitude, at the equator, about 2.5 times greater than that of $S_2^2(T)$ at the equator. This illustrates in numerical terms, for the main parts of $S(p)$, the contrast remarked by Kelvin between their ratio, and that of $S_1(T)$ and $S_2(T)$.

2S.4C. $S_3(p)$

This 8-hourly variation was studied by Hann (1918), Schmidt (1919), Bartels and Kertz (1952) and others. Its geographical distribution is rather regular, and it has a notable seasonal reversal of phase from summer to winter in each hemisphere, being approximately antisymmetrical relative to the equator. Bartels and Kertz expressed it thus:
$$0.293P_4^3(\theta)\sin(3t+\sigma_3)\text{ mb}; \quad P_4^3=(35/8)^{1/2}\sin^3\theta\cos\theta.$$

In season j, $\sigma_3 = 149°$, in season d it is $335°$. This phase reversal comes from a corresponding phase reversal of $S_3(T)$, the cause of $S_3(p)$; see Bartels (1927; also p. 675, Bartels and Kertz, 1952) for Potsdam month-to-month dial polygons for $S_3(P)$ and $S_3(T)$.

Siebert (1957), on the basis of Hann's data for January and July only, modified the factor $0.293P_4^3$ in the above expression for the seasonally reversing part of $S_3(p)$ to

$$0.269P_4^3 - 0.014P_6^3,$$

and gave σ_3 for July as $175°$ and for January as $355°$: he found also a small part that remains constant throughout the year, with amplitude

$$0.075P_3^3 + 0.044P_4^3 + 0.039P_5^3$$

and phase $10°$.

2S.4D. $S_4(p)$

The component $S_4(p)$ has been discussed by Pramanik (1926) and Kertz (1956c, 1957, 1959). It is small, and its geographical variations are greater than those of $S_2(p)$ and $S_3(p)$, but it has more regularity than $S_1(p)$. Kertz (1956c) discussed $S_4(p)$ in relation to $S_4(T)$, extending Pramanik's study of the annual mean results to include also the seasonal (j, e, d) results. Using monthly mean data for $S_4(p)$ for 22 stations, regrettably not well distributed, given by Angot (1889), each based on many years' record, he analyzed them in the manner outlined in Section 2S.3, and obtained the ellipses of the annual and semi-annual variations of A_4, B_4 on a harmonic dial. Almost all were narrow, and could be approximated by straight lines. The ends of the annual ellipse major axis were the points for summer and winter, for the semi-annual ellipse they were the points for the seasons e and $(j+d)/2$. Hence the seasonal mean values A_4, B_4 for j, e, d indicate the variation of $S_4(p)$ throughout the year fairly well. Kertz (1956c) gave a table of their values for 62 stations, from analyses by Hann, Angot and Chapman; many of these are for the stations and periods indicated in the paper on $L_2(p)$ by Chapman and Tschu (1948).

Using these data, Kertz (1956c) gave the following amplitudes and phases for the 4 main terms in the spherical harmonic expression of the geographical distribution of $S_4(p)$.

TABLE 2S.6

$$S_4(p) = \sum_{m=4}^{7} s_m{}^4 P_m{}^4 \sin(4t + \sigma_m{}^4)$$

for the seasons j, e, d and for the year y: the unit here used is the microbar

Amplitude $s_m{}^4$ (μb)				Phase $\sigma_m{}^4$			
m	j	e	d	y	j	e	d	y
4	19.1	17.4	7.9	13.4	180°	173°	244°	187°
5	24.2	27.9	98.1	35.1	359	204	207	213
6	26.8	34.7	24.4	15.5	222	336	215	260
7	22.8	13.9	28.2	10.8	331	252	191	248

2S.5. The Daily Variation of Air Temperature T

There is a great volume of data for the air temperature T, mainly for its value T_0 at the surface. Its daily variation $S(T_0)$ has not been discussed and analyzed so fully as has been done for $S(p)$. Haurwitz (1965) has discussed the diurnal component $S_1(T)$, the annual mean data for which he has analyzed (Haurwitz, 1962). From his analysis he has deduced, partly in combination with theory, the expression of the main wave $S_1^1(T_0)$ in terms of spherical harmonic functions (as he did for $S_2(p)$; cf. Section 2S.4A) for the terms of degree $|s| + i$ equal to 1, 2, 3. Theoretically such expressions have also been deduced, for the equinoctial season, by Siebert (1961) on the basis of absorption of solar radiation, and by Kertz (1957), who took account also of mass exchange by turbulence. Table 2S.7 gives their results; the main term proportional to P_1^1 is not greatly larger than the terms in P_2^1 (asymmetrical to the equator) and P_3^1. "As shown by Laplace (see Lamb, 1932, pp. 341–2), no changes in the elevation of the free surface of an ocean of uniform depth covering the whole earth will occur if the generating force is of the form P_2^1. We may therefore surmise that the corresponding term in the diurnal temperature oscillation will not cause an appreciable pressure oscillation of this form" (Haurwitz, 1965).

These ground data, however, need to be considered along with data for $S(T)$ at higher levels (Section 2S.5), in considering the (certainly thermal) generation of $S_1(p)$.

Haurwitz and Möller (1955) analyzed the annual mean data for $S_1(T)$ as follows:

TABLE 2S.7

Harmonic coefficients of the diurnal surface temperature wave

Wave Type	$|s| + i$	Haurwitz, Möller		Kertz		Siebert	
		Amp.	Phase const.	Amp.	Phase const.	Amp.	Phase const.
	1	1.007°C	232°	0.748°C	225°	0.158°C	180°
$S_1^1(T_0)$	2	0.647	232	0.439	225	–	–
	3	0.502	238	0.083	225	0.020	0

Studying the $S(T)$ curves for many stations they distinguished three types, polar, temperate and tropical; in any one of these areas, $S(T)$ was similar in type, but its range A differed from station to station. They determined standard polar, temperate and tropical forms of $S(T)$, here denoted by S_p, S_{te}, S_{tr}, of range 1 °C; for each of these they gave bihourly values of the deviation from the mean. They divided the earth's surface into a set of areas bounded by meridians 20° apart and latitude circles 10° apart; for each of these they determined the fraction k covered by land. They read the range A from a chart given by Shaw (1936), and took $S(T)$ for each such area to be given by kAS, where S signifies S_p, S_{te} or S_{tr} according to the situation: the semidiurnal component of the variation, for all the areas, was their basis for a spherical harmonic expression of S_2, made along the lines already described in connection with $S_2(p)$. The expression included a zonal standing wave $S_2^0(T)$, expressed as follows, the dominant term being P_3.

$$S_2^0(T) = 0.024 \sin(2t_u + 219°) + 0.076\, P_1 \sin(2t_u + 194°)$$
$$+ 0.040 P_2 \sin(2t_u + 214°) + 0.112\, P_3 \sin(2t_u - 1°) + 0.104\, P_4 \sin(2t_u + 56°).$$

For the part of the wave that follows the sun, $S_2^2(T)$, they modified their material by taking k to be the same (a mean value) over each belt of latitude, independent of longitude. This led to the following expression:

$$S_2^2(T) = 0.301\, P_2^2 \sin(2t + 65°) + 0.368\, P_3^2 \sin(2t + 65°)$$
$$+ 0.113\, P_4^2 \sin(2t + 68°).$$

Here the term P_3^2 asymmetrical with respect to the equator exceeds the other main term P_2^2, the reason being the great difference in the land fractions on the two sides of the equator.

For the P_2^2 term (only) they revised the calculation by taking account of the variation of k in each belt of latitude. This led to the conception of a semidiurnal temperature variation whose phase varies with Greenwich time. This complicated picture would be still more complicated if the P_3^2 term were taken into account.

These determinations of $S_2(T)$ and $S_2^2(T)$ were discussed in connection with the thermal excitation of the corresponding p variations. But further consideration of this problem is required, to take account of the influence of $S_2(T)$ at other levels, such as in the ozone layer.

Siebert (1957), using January and July data for $S_3(T)$, gave the following expression for the amplitude of the seasonally reversing part:

$$0.018 \ P_3^3 + 0.075 \ P_4^3 + 0.064 \ P_5^3 + 0.057 \ P_6^3 + 0.050 \ P_7^3,$$

with phase 45° in July and 225° in January: and the following for the amplitude of the unchanging part, with phase 12°:

$$0.067 \ P_3^3 + 0.030 \ P_4^3 + 0.014 \ P_5^3.$$

The theoretical study of $S(T)$ made by Kertz (1956a) led to results expressing various components in terms of spherical harmonics. Kertz (1956b) discussed the nature and geographical distribution of $S_4(T)$ in relation to $S_4(p)$, and also compared his results with the observational data (Pramanik, 1926) for continental stations (only). Kertz took $S(T)$ to be zero, to a first approximation, over the water. The theoretical amplitudes were found to be too small by a factor of order 2, and phase differences also were found for some of the P_m^4 terms. No clear conclusions were reached regarding the relation between $S_4(T)$ and $S_4(p)$, and further study of the observed data and of their seasonal variations was urged.

2S.6. The Daily Wind Variation $S(V)$

The daily wind changes have been studied much less than $S(p)$ and $S(T)$. They depend greatly on topography and weather (e.g., land and sea breezes near coasts). The observations also depend much on such local influences as the height and situation of the anemometers, which have far less effect on the barometer. These local effects on wind are described in meteorological textbooks.

The world-wide distribution of the wind has some notable planetary features, such as the trade and counter-trade winds, but we know little about the world-wide distribution of the daily variation $S(V)$. Möller (1940) attempted, on the basis of a 10-year series of Potsdam wind data, to distinguish between a convective part due to local heating and cooling, and a pressure gradient part.

In the process of determining the *lunar* tidal wind variation $L(V)$, for Mauritius (Chapman, 1949) and for Uppsala and Hongkong (Haurwitz and Cowley, 1969), values of $S_2(V)$ were obtained, as follows, for $S_2(u)$, $S_2(v)$ – see 2L.13; here u and v denote respectively the southward and eastward wind components of V (in Chapter 3 these are referred to as northerly and westerly); the unit is 1 cm/sec.

Mauritius	*Hongkong*	*Uppsala*
20°.1 S; 16 years	23°.3 N; 67 years	59°.9 N; 84 years
$S_2(u)$: 13.4 sin $(2t + 263°)$;	18.1 cm/sec ± 0.4; 199°.4;	2.3 cm/sec ± 0.4; 341°
$S_2(v)$: 27.6 sin $(2t + 92°)$;	13.0 cm/sec ± 0.4; 288.3;	8.4 cm/sec ± 0.3; 51°.

Figure 2L.9, lower part, p. 103, gives the harmonic dial vectors for $S_2(-u)$, $S_2(v)$ at Mauritius. Figure 2L.10, lower part, p. 104, shows how the $S_2(V)$ vector changes during the day – the curve is described twice daily, before and after noon. The same curve describes the path of a particle that moves solely according to $S_2(V)$; for this interpretation the scale on the left of the origin is to be used, and all the hour numbers are to be increased by 3. Thus the total excursion of the particle on either side of its mean position is 23 km.

It is of interest to compare the value of $S_2(\mathbf{V})$ obtained from the observations, with the value calculated from $S_2(p)$, for an atmosphere on a rotating earth, but ignoring friction. The equations of motion are as follows; a denotes the earth's radius, ω its angular velocity:

$$\frac{\partial u}{\partial t} - 2\omega v \cos \theta = -\frac{1}{\varrho} \frac{\partial p}{a \partial \theta}$$

$$\frac{\partial v}{\partial t} + 2\omega u \cos \theta = -\frac{1}{\varrho} \frac{\partial p}{a \sin \theta \, \partial \phi}$$

We consider only the main term in $S_2(p)$, given in 2S.4a, namely

$$p_S \sin^3 \theta \sin (4\pi t_u/T_S + 2\phi + \sigma), \quad p_S = 1.16 \text{ mb}, \sigma = 158°.$$

Here t_u is taken to be expressed in seconds. Thus in connection with S_2,

$$\frac{\partial}{\partial t} = \frac{2\pi}{T_S} \frac{\partial}{\partial \phi}.$$

It is readily verified that the solution of the equation for $S_2(\mathbf{V})$ is

$$S_2(u) = 2.5 \, C_S \cos \theta \sin (2t + \sigma + 90°).$$
$$S_2(v) = C_S (1 + 1.5 \cos^2 \theta) \sin (2t + \sigma + 180°).$$

where

$$C_S = p_S/\varrho a\omega = 20.0 \text{ cm/sec}.$$

Thus the 'theoretical' phase for $S_2(u)$ should be 248° in northern latitudes, and 68° in southern latitudes; for $S_2(v)$ it should be 338° in all latitudes. These phase predictions do not agree well with the determined results given above. The amplitudes are still more discordant; for $S_2(u)$ they are 17.2, 19.8 and 40.3 cm/sec, and for $S_2(v)$ they are 23.5, 24.7 and 39.4, respectively for Mauritius, Hongkong and Uppsala. The theoretical amplitudes increase with latitude; instead the observational values decrease. For Mauritius the amplitude agreement is fairly good, but with increasing latitude it changes to disagreement, pronouncedly so for Uppsala.

As remarked by Haurwitz and Cowley (1968), the angle α, measured anticlockwise, towards the east, between the southerly direction and the wind vector, is given by $\tan \alpha = v/u$. Hence differentiation with respect to time gives $d\alpha/dt$ by the equation $(u \sec \alpha)^1 \, d\alpha/dt = 2u_0 v_0 \sin(\sigma_u - \sigma_v)$, where u_0, v_0 denote the amplitudes, and σ_u, σ_v the phases, of $S_2(u)$ and $S_2(v)$. Thus, if $\sin(\sigma_u - \sigma_v)$ is positive, the wind vector rotates anticlockwise, or clockwise if it is negative. For Mauritius $\sin(\sigma_u - \sigma_v)$ is positive; hence, as shown by Figure 2L.10, its $S_2(\mathbf{V})$ vector rotates anticlockwise. In this respect, but not as regards the magnitude of $\sigma_u - \sigma_v$, $S_2(\mathbf{V})$ at Mauritius agrees fairly well with the theory. At the two northern stations the phase differences, but not the phases, agree fairly well, in sign and magnitude, with the theory. It remains to be considered whether the neglect of friction, in the equations of motion, can account for these discrepancies with theory; certainly near the ground friction must modify these systematic winds.

2S.7. Atmospheric Daily Changes above Ground Level

The great bulk of tidal and thermotidal analyses have been made for surface data. Obviously this is because such data are most readily available. However, at the surface, daily variations (at least those on a global scale) usually constitute only a small fraction of the meteorologically interesting variation. As pointed out in 1.11, it is only higher in the atmosphere that daily variations comprise an important and eventually dominant part of the total fields. Above the ground, daily variations are not only scientifically interesting, they are also major components of the physics of the atmosphere. It is, therefore, regrettable that above-ground data are both rarer and more expensive to obtain than surface data. Nevertheless, in view of the large relative and absolute magnitude of daily variations in the upper atmosphere, less material is necessary to isolate the various oscillations, and over the last 15 years or so, appreciable progress has been made in this area.

2S.7A. DAILY VARIATIONS BETWEEN THE GROUND AND 30 KM

Data at these heights are available from balloons for both wind and temperature – though data for the former appear more common and reliable. Data from many stations are available every 12 hrs, while a smaller number of stations take soundings every 6 hrs. Data are taken at standard hours, which, at present, are 0000 and 1200 GMT for stations taking soundings every 12 hrs, and 0000, 0600, 1200 and 1800 GMT for stations taking soundings every 6 hrs. Prior to 1 June 1957, the standard hours were 0300 and 1500 GMT and 0300, 0900, 1500 and 2100 respectively. Thus, there are available for many stations long term means of the wind at 6-hour intervals, and for fewer stations at 3-hour intervals. This situation existed somewhat earlier in Great Britain, where in the early 1950's there existed nearby stations taking soundings at 0000, 0600, 1200 and 1800, and at 0300, 0900, 1500 and 2100. Such data give eight values for the wind equispaced over a day. These values can be harmonically analysed in the manner described in Section 2S.2, to give the first four harmonics. This was first done by Johnson (1955) for data from 150 mb–100 mb over England. Data for one year were used. The deviation of the annual mean wind components for stated hours from the overall means is shown in Table 2S.8, taken from Johnson's paper.

TABLE 2S.8

Deviation of annual mean wind-components for stated hours from the overall means (Johnson, 1955)

Time GMT	03	06	09	12	15	18	21	24
u (cm/sec)	− 0.5	+27.8	+12.9	−53.0	−48.9	+20.6	+36.6	+ 5.7
v (cm/sec)	22.1	− 2.6	−47.4	−36.1	23.7	28.3	1.5	10.3

The harmonic analysis of the data in Table 2S.8 yields

$$S(u) = \{27.8 \sin(\theta + 92°\,14') - 34.5 \sin(2\theta + 74°24')$$
$$+ 3.1 \sin(3\theta + 80°32') - 0.13 \sin(4\theta - 30°)\} \text{ cm/sec}, \tag{1}$$

$$S(v) = \{26.2 \sin(\theta + 144°24') + 26.2 \sin(2\theta + 57')$$
$$- 3.6 \sin(3\theta - 26°34') + 0.0 \sin(4\theta)\} \text{ cm/sec} \tag{2}$$

where θ goes from 0 to 360° as time goes from 0 to 24 hrs GMT. The most notable features of Equations (1) and (2) are the dominance of the diurnal and semidiurnal components, and their almost equal amplitude. Johnson (1955) performed a careful statistical analysis of his data, showing it to be significant. He did not, however, calculate probable error circles. A perusal of his monthly means suggests that the error circles would have a radius of about 5 cm/sec. Subsequently, similar analyses were done on collections of data from Washington, D.C. (Harris, 1959) and from Terceira, Azores (38°44'N, 27°4'W.) (Harris, Finger and Teweles, 1962). Analyses for eight more stations between 30°N and 76°N were made by Harris, Finger and Teweles (1966), whose results, given in Table 2S.9, are typical of those they obtained at all stations. Amplitudes are of order 30 cm/sec, and phases are typically such that the diurnal component of the wind blew northward at about local noon; maximum westerly flow occurred about 6 hours later. A mild decrease in amplitude from 30°N to 76°N was noted for both diurnal and semidiurnal oscillations. There were also some stations (primarily continental stations) where the diurnal oscillation had a distinct maximum in amplitude near the ground (ca. 70 cm/sec); at these stations the amplitude of the diurnal oscillation was greater (at most tropospheric levels) than at the other stations. In general, however, the picture of the daily wind variations presented by Harris, Finger and Teweles (1966) is fairly simple. A study of the daily variations over the central United States by Hering and Borden (1962), using 6-hourly data for summer, 1958, showed a much more complicated picture. Their data indicated distinct maxima (ca. 1 m/s) in the diurnal wind oscillation amplitude at 1 km, 5 km and 12 km. These maxima appeared to be associated with surface topography; Wallace and Hartranft (1969) confirmed this by an extensive study. They used the previously established fact, that the daily variation of wind consists predominantly of the diurnal and semidiurnal components, to conclude that the long term average of the differences between wind data taken at 1200 UT and 0000 UT should yield the diurnal component of the wind at 1200 UT. Moreover, most radiosonde stations take their soundings simultaneously in universal time. Thus, if one draws the twelve hour wind difference vectors at a given level for many stations on a map, one will have a picture of the general circulation due to the diurnal wind oscillation at that level. If the diurnal oscillation were merely travelling westward with the sun, this circulation would consist in only zonal wave number one (i.e., longitude dependence would be given by $\sin(\phi + \alpha)$). Instead, at low levels there were circulation gyres associated with major topographical features such as mountain ranges. The influence of these features was evident as high as 200 mb, but diminished at still higher levels. Above 200 mb, however, there remained gyres associated with the continents and oceans. By 15 mb the simple zonal wave number one

TABLE 2S.9

Diurnal and semidiurnal variations of the eastward and northward components of the wind at Terceira, Azores. Annual mean values of the amplitude, s, in cm/sec, and phase, σ, in degrees. P.E. is the radius of the probable error circle of the annual means. Time of maximum wind is related to σ by the expression $t_{max.} = (450° - \sigma)/15n$, where $n = 1$ for diurnal variations, and $n = 2$ for semidiurnal variations. After Harris, Finger and Teweles (1962)

Mean pressure, mb	Variation of eastward wind component						Variation of northward wind component					
	Diurnal			Semidiurnal			Diurnal			Semidiurnal		
	s_1	σ_1	P.E.	s_2	σ_2	P.E.	s_1	σ_1	P.E.	s_2	σ_2	P.E.
Ground	2	75°	6	8	324°	7	7	341°	8	21	52°	6
1000	4	115	9	12	298	6	6	337	6	17	52	6
950	2	154	8	14	317	8	21	271	11	22	35	6
900	2	248	3	19	292	6	32	272	9	23	31	8
850	8	257	11	14	266	7	25	256	13	23	14	8
800	4	322	10	22	313	8	18	265	11	31	359	8
750	22	145	14	22	278	10	16	251	15	29	12	14
700	5	304	11	18	304	8	9	4	11	22	33	9
650	4	255	11	20	292	5	13	318	9	23	50	8
600	8	63	12	20	272	6	12	281	11	16	1	7
550	20	159	10	31	327	9	18	249	13	16	7	10
500	20	124	9	25	276	10	15	317	15	15	63	9
450	17	76	11	26	295	10	22	295	16	20	346	10
400	18	1	8	28	291	10	14	342	17	10	317	12
350	19	258	12	42	291	9	13	257	19	16	319	9
300	24	193	18	51	292	14	8	247	21	14	4	13
250	52	177	16	26	245	12	52	267	17	8	285	12
200	56	153	15	46	267	11	18	238	14	15	338	12
175	13	164	13	37	278	14	34	241	13	25	339	10
150	16	186	11	39	300	11	18	185	14	52	18	9
125	14	127	8	29	242	9	5	241	7	23	4	8
100	27	112	10	55	280	10	34	153	11	28	19	6
80	31	111	11	37	280	9	21	194	9	28	21	7
60	34	109	11	36	262	8	40	196	9	27	5	10
50	19	132	9	41	256	8	34	236	10	42	356	7
40	6	96	11	44	263	12	27	235	11	49	4	8
30	23	181	11	67	280	13	21	221	10	65	17	11
20	30	147	12	62	295	12	64	235	13	60	36	14
15	25	114	23	91	303	20	66	238	14	61	30	10
10	–	–	–	–	–	–	–	–	–	–	–	–

pattern appears to be dominating. Wallace and Hartranft also showed that topographical effects are small at high latitudes and during winter; they occur primarily during the summer at temperature and tropical latitudes. At tropical latitudes topographical disturbances penetrate to greater altitudes than at higher latitudes. These features are seen in Figures 2S.9–11.

In Figure 2S.9 we also see vectors representing wind differences between 0300 and 1500 obtained from data obtained before June 1957.

2S.7B. DAILY VARIATIONS FROM 30 KM–60 KM

Since 1959, regular soundings of the wind between 30 and 60 km have been made at a small number of stations using meteorological rockets and a variety of techniques.

Fig. 2S.9.   Annual average wind differences 0000–1200 GMT (solid) and 0300–1500 GMT (dashed) at 700 mb, plotted in vector form. The length scale is given in the figure. After Wallace and Hartranft (1969).

Unfortunately, most soundings were made near local noon. Thus, it appeared that special series of closely spaced soundings would be needed in order to detect daily variations. Sequences of 2-hourly soundings were taken over 24 hours over Eglin Air Force Base on 9–10 May 1961 (Lenhard, 1963), and over White Sands Missile Range on 7–8 February 1964 and on 21–22 November 1964 (Miers, 1965). Another sequence of 16 soundings over 51 hours was taken at White Sands Missile Range between 30 June and 2 July 1965 (Beyers, Miers and Reed, 1966). It is clear that such limited data hardly allow meaningful error analyses. However, all the experiments suggested the presence of a strong daily variation above 40 km. This is seen in Figure 2S.12 taken from Beyers, Miers and Reed (1966), showing the deviation of the meridional wind from its mean as a function of time. There can be little doubt that daily variations dominate the deviation. From Figure 2S.12 we estimate the amplitude of the diurnal wind oscillation to be about 8 m/s at 46 km. This amplitude is sufficiently great to suggest that our earlier statement that routine soundings were inadequate for determining

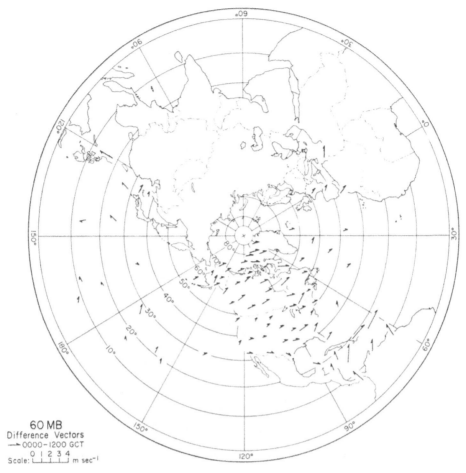

Fig. 2S.10.   Annual average wind differences 0000–1200 GMT at 60 mb plotted in vector form. The
length scale is given in the figure. After Wallace and Hartranft (1969).

daily variations be reconsidered. If one divides the day into 12 two-hour segments,
then we find that there are two stations (White Sands and Cape Kennedy – both near
30 °N) where there are several soundings in each time segment. One could, therefore,
calculate mean wind profiles for each time segment, and harmonically analyse these
means at each altitude. Unfortunately, the soundings for each time segment are often
irregularly distributed over the year, and there are problems of trend removal. For
zonal wind, we know that there are variations between winter and summer of order
100 m/s. In addition the winter mesosphere is disturbed by large scale planetary waves
with periods of order 5 days. These affect both the zonal and meridional components
of the wind. However, the summer mesosphere appears to be relatively free from these
large scale disturbances, and the mean meridional flow in the summer mesosphere
appears to be very small. Thus, it appears possible to use routine soundings at
White Sands and Cape Kennedy during the summer to determine the daily variation

Fig. 2S.11. Annual average wind differences 0000–1200 GMT at 15 mb plotted in vector from. The length scale is given in the Figure. After Wallace and Hartranft (1969).

in the meridional (i.e., northerly) wind. This was done by Reed, McKenzie and Vyverberg (1966a), whose results are entirely consistent with the previously mentioned analyses. Figures 2S.13 and 2S.14 show vertical distributions of amplitude and phase of the diurnal oscillation in northerly wind from various analyses. The amplitude is small (less than 1 m/s) below about 37 km (consistent with findings cited in Section 2S.7A), rising rapidly to a maximum of about 8 m/s near 50 km and decreasing again above 50 km. There is some evidence of a resumed increase in amplitude above 60 km. The phase of the oscillation is such that maximum northerly wind occurs near 0000 local time at the level of maximum amplitude. Reed (1967) also attempted to use routine data to determine the solar semidiurnal oscillation at White Sands and at Ascension Island (8 °S). His results for White Sands are shown in Figure 2S.15. The determination at 30 km appears consistent with the results cited in Section 2S.7A. The amplitude of the semidiurnal oscillation remains small ($\sim$1 m/s) until 50 km, where it begins to increase to values comparable with the amplitude of the diurnal oscillation. There is clearly a 180° shift of phase between the balloon levels and the region above 50 km. Given the small amplitude of the semidiurnal variation below 50 km, however, it is

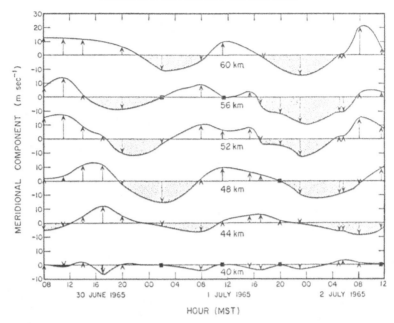

Fig. 2S.12. Meridional wind components u in m/sec. averaged over 4 km layers centered at 40, 44
48, 52, 56 and 60 km. Positive values indicate a south to north flow. After Beyers,
Miers and Reed (1966).

doubtful whether the level at which the phase shift occurs can be reliably determined
with currently available data.

The ease of determining diurnal oscillations in the meridional wind during summer
led Reed, Oard and Sieminski (1969) to analyse data from regions with fewer soundings. First, they combined soundings from stations at nearby latitudes (namely, Fort
Churchill (59 °N) and Fort Greely (64 °N); Green River (39 °N), Wallops Island (38 °N)
and Point Mugu (34 °N); White Sands (32 °N) and Cape Kennedy (28 °N); Barking
Sands (22 °N), Grand Fork (21 °N) and Antigua (17 °N); Ascension Island (8 °S)).
The resulting data for each group of stations was analysed at one kilometer altitude
intervals by fitting diurnal and semidiurnal harmonics to the irregularly-spaced data
points by the method of least squares. Each datum was weighted equally. Since
soundings at the stations considered were concentrated primarily between the hours
of 0800 and 1600 local time, the analysis amounted, in essence, to analysing about
8–10 hours of data for diurnal oscillations. This method is adequate for determining
amplitudes and phases if one is certain *a priori* that the meridional wind variations
are due entirely to diurnal and semidiurnal oscillations. What happens under more
general circumstances is analysed in Section 2S.7E. In the present case Reed (private
communication) has stated that more conventional methods of analysis, giving equal
weight to each time interval, yielded basically similar results. Thus, the results of
Reed *et al.* (1969) are at least likely to be meaningful. Some examples of their results
are given in Figures 2S.16 and 2S.17, giving the distribution with height of the ampli-

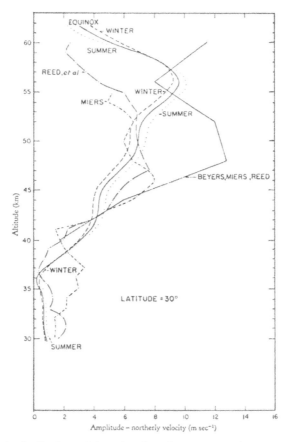

Fig. 2S.13. Altitude distributions of the solar diurnal component of the southward wind u at 30°
latitude based on various observations. Also shown are theoretical distributions for winter, equinoctial
and summer conditions. After Lindzen (1967a).

tude and phase of the diurnal oscillation at stations near 61 °N and near 20 °N. The
most notable difference in the oscillation at these two latitudes is in the variation of
phase with height: pronounced at 20 °N, it is virtually non-existent at 61 °N. Figures
2S.13, 14, 16 and 17 show curves derived theoretically. These are discussed in Section 3.7.

In addition to wind measurements, meteorological rockets are also used for
temperature measurements (using thermistor beads). Sequences of temperature
soundings over White Sands appear to show day-night temperature differences at
45 km of order 20 °C (Beyers and Miers, 1965) – much larger than expected on the
basis of radiative considerations (Pressman, 1955; Johnson, 1953). Much controversy
was stimulated by the large claimed daily temperature variation, and it now appears
that the changes observed were largely due to radiation errors (Hyson, 1968). Finger
and McInturff (1968), carefully analysing balloon data in such a manner as to eliminate radiation errors, and using balloon soundings reaching 36 km, found that
day-night temperature differences at 36 km are indeed much smaller than indicated

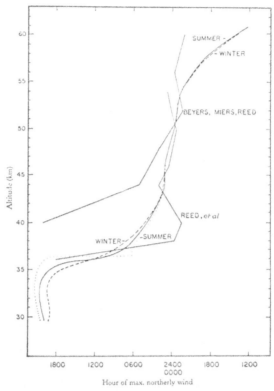

Fig. 2S.14. Altitude distributions of the phase of the solar diurnal component of the southward wind u at 30° latitude based on various observations. Also shown are theoretical distributions for winter, equinoctial and summer conditions. After Lindzen (1967a).

by rocket data. Finger and McInturff's analysis suggests an amplitude for the diurnal temperature oscillation at 36 km varying from 1 °C at 40 °N to 0.5 °C at 70 °N, with maximum temperature occurring near 1800 hrs local time.

2S.7C. DAILY VARIATIONS FROM 80–120 KM

Rockets capable of sounding the atmosphere above 60 km are, for the time being, too costly for routine use. While much useful rocket information is available for altitudes above 60 km, it is neither as plentiful nor as regular as the data from below 60 km. Thus, most of our data for this region comes from natural phenomena which make possible the indirect sensing of upper atmospheric winds from the ground.

The most important measurement technique at present consists in observing by means of doppler radar the motion of ionized trails left by meteors disintegrating on traversing the atmosphere. Such disintegrations occur mainly between 80 and 110 km. At these altitudes the collision frequency for ions with neutral air molecules is much greater than the ionic gyromagnetic frequency. Thus, the ions move with the neutral air and serve as a tracer of neutral air motions. The first analyses of upper atmospheric

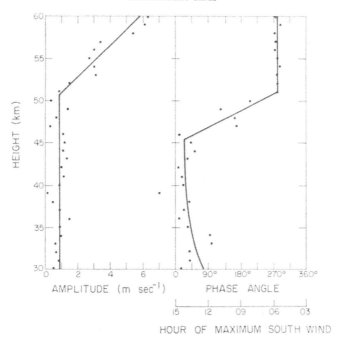

Fig. 2S.15.   Phase and amplitude of the semidiurnal variation of the meridional wind component u at 30°N, based on data for White Sands (32.4N) and Cape Kennedy (28.5N). After Reed (1967).

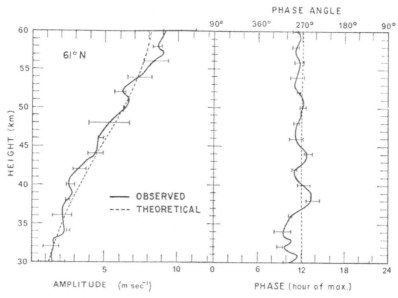

Fig. 2S.16.   Amplitude and phase of the diurnal variation of the meridional wind component u at 61°N. Phase angle, in accordance with the usual convention, gives the degrees in advance of the origin (chosen as midnight) at which the sine curve crosses from $-$ to $+$. After Reed, Oard and Sieminski (1969).

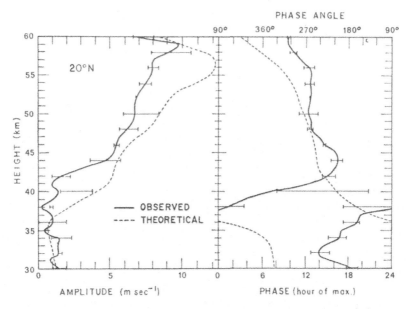

Fig. 2S.17. Amplitude and phase of the diurnal variation of the meridional wind component u
at 20°N. After Reed, Oard and Sieminski (1969).

daily variations were made by Greenhow and Neufeld (1961) at Jodrell Bank (53 °N)
and by Elford (1959) at Adelaide, Australia (35 °S). Greenhow and Neufeld used data
for about 100 days scattered through the years 1953–58, while Elford used data for
about 400 days from 1952 until 1955. Their observed winds were, in fact, averages
over 80–100 km. Haurwitz (1962a) reviewed tidal phenomena in the upper atmosphere, and prepared convenient displays (1964) of the results obtained by Greenhow
and Neufeld, and by Elford. These are shown here in Figures 2S.18 and 2S.19. From
Figure 2S.18 we see that the diurnal contribution to the average wind between 80 and
100 km above Jodrell Bank is typically only about 5 m/s, and the semidiurnal contribution is typically 13 m/s (the numbers refer to oscillation amplitudes). From
Figure 2S.19 we see that diurnal amplitudes above Adelaide are typically more than
20 m/s, and the semidiurnal amplitudes there are only of order 10 m/s. In both figures,
E refers to equinoctial months (March, April, September and October), J to summer
months (May, June, July and August), and D to winter months (November, December, January and February). An important part of Haurwitz's presentation is the
estimation of probable error circles. A meaningful determination generally requires
that the magnitude of a quantity be more than three times the radius of the probable
error circle. By this criterion, it is clear that several of the seasonal means shown in
Figures 2S.18 and 2S.19 are not reliably determined.

Table 2S.10 (p. 61) (Haurwitz, 1964), indicates how often, among the data considered, the mean wind, in each component, is greater or less than the amplitude
(s_1 or s_2) of the diurnal and semidiurnal wind variations. It shows that between 80
and 100 km the daily oscillations are a major part of the general circulation there.

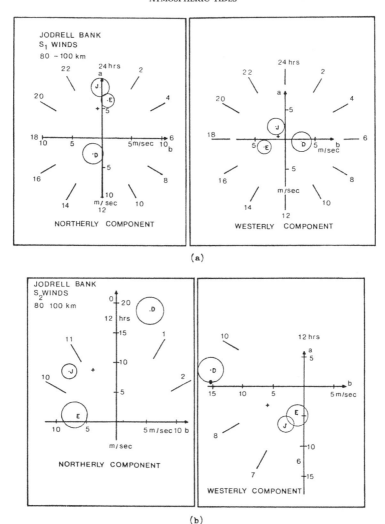

Fig. 2S.18. Harmonic dials for the mean northerly and westerly components of (a) the diurnal and (b) the semidiurnal wind variations at Jodrell Bank at 80–100 km. Crosses indicate annual mean values, dots seasonal values. Circles show probable errors of the seasonal means.
From Haurwitz (1964).

More recently, radio meteor techniques capable of observing hourly winds with height resolutions of order 1 km have been developed (Revah, Spizzichino and Massebeuf, 1967). Analyses of early observations over Garchy, France (47°N) show the presence of a semidiurnal oscillation whose amplitude varies from about 10 m/s at 80 km to 40 m/s at 100 km. The phase increases upward at a rate appropriate to a vertical wavelength of about 90 km, though in some months the phase increase was more rapid.

The daily variations in wind at altitudes from about 110 to 120 km, as inferred from dynamo calculations, are somewhat difficult to interpret. Such calculations have

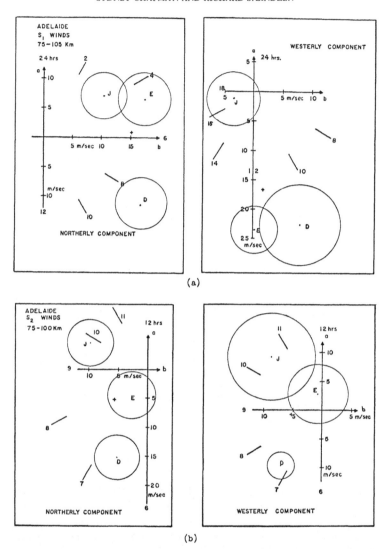

Fig. 2S.19.   Harmonic dials for the mean northerly and westerly components of (a) the diurnal and (b) the semidiurnal wind variations at Adelaide, Australia, at 80–100 km. Crosses indicate annual mean values, dots seasonal values. Circles show probable errors of the seasonal means.
From Haurwitz (1964).

hitherto assumed that winds are independent of altitude (which, as we have seen, is not the case below 110 km). Using electrical conductivities calculated from observations of electron distributions with height and their time variations, one can calculate the time varying 'slab' winds giving rise to the observed quiet day daily variations in the geomagnetic field. Such calculations have been made by Maeda (1955) and Kato (1956). Some complications and uncertainties involved in these calculations are discussed by Hines (1963) and Price (1969). Kato's results are shown in Figure 2S.20.

TABLE 2S.10

Number of cases when the mean wind is larger or smaller than the amplitude of the first (S_1) and second (S_2) harmonic of each wind component. From Haurwitz (1964).

	Jodrell Bank		Adelaide	
	NS	EW	NS	EW
$\bar{v} > S_1$	33	58	5	16
$\bar{v} < S_1$	61	34	32	21
$\bar{v} > S_2$	15	41	9	23
$\bar{v} < S_2$	81	57	28	14

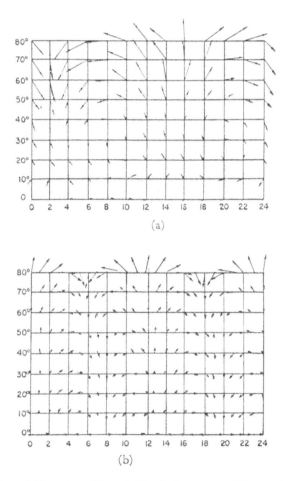

Fig. 2S.20. The diurnal (a) and semidiurnal (b) wind patterns of the northern hemisphere, as deduced by Kato (1956) from the average quiet-day daily magnetic variations. The sides of each elementary square represent a wind speed of 50 m/sec. From Hines (1963).

They suggest that the diurnal circulation is stronger than the semidiurnal circulation, and that both are larger at high latitudes than at low latitudes.

Rocket soundings of the atmosphere above 80 km are still too infrequent (and sometimes unreliable) to provide much information concerning daily variations. Wind measurements between 90 and 120 km can be made by visually observing vapor trails released by rocket. Unfortunately, most such measurements have used sodium vapor, which can only be seen in twilight. Nevertheless, such soundings (a total of 29 soundings were made by Manring, Bedinger, Knaflich and Layzer (1964) between August 1959 and May 1963 from Wallops Island, Virginia and from Sardinia – both near 38 °N) may be grouped into dawn soundings and dusk soundings. If one assumes that the wind is primarily due to a prevailing wind plus diurnal and semidiurnal oscillations, then the difference between dawn and dusk soundings should be indicative of the diurnal range (more precisely, it should be a lower bound). Similarly, the average of dawn and dusk soundings should be due to a combination of the prevailing wind and the semidiurnal oscillation. An analysis of the differences and means of the averages of dusk and dawn soundings has been made by Hines (1966). Figure 2S.21

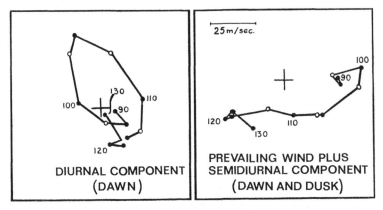

Fig. 2S.21.   Vectograms showing the diurnal tide at dawn and the prevailing wind plus the semidiurnal tide at its dawndusk phase, as functions of height. After Hines (1966).

shows his findings. The amplitude of the diurnal oscillation appears to increase from about 10 m/s at 90 km to about 50 m/s at 105 km, and to decrease again above 105 km. The wind vector rotates about 360° in 22 km, indicating a wave with a vertical wavelength of about 22 km. New vapors permit the observation of chemi-luminescent trails throughout the night (Rosenberg and Edwards, 1964). Though soundings using such vapors will not provide data over a full day, tidal analyses may still be possible along the lines described in 2S.7B and analysed in 2S.7E.

2S.7D. DAILY VARIATIONS IN THE THERMOSPHERE

Artificial satellites regularly travel through the earth's atmosphere above 200 km. Air exerts a drag on the satellite proportional to the air density, and the drag modifies the satellite's orbit in a predictable manner. Given the relation between drag and air density, and precise measurements of satellite orbits, one may obtain extensive data for atmospheric density. The method is described in detail by Jacchia (1963). Extensive measurements (Harris and Priester, 1965) show an extremely strong daily variation in density at all levels above 200 km. Density varies by a factor of five during a day at 360 km, maximum densities occur near 1400 hours local time.

2S.7E. ANALYSIS OF DATA COVERING ONLY A FRACTION OF A DAY

It is clear from 2S.7b and 2S.7c that for some time to come much of the available upper atmosphere data will cover only part of the day (near noon from 30–60 km, night from 100–120 km). The question arises as to whether such data can be adequate for determining daily variations. Clearly, if data are available at 5 different times, at least, within the fractional segment of a day, and if we know, *a priori*, that the wind consists *only* in a prevailing wind plus diurnal and semidiurnal oscillations, then the data will be adequate; we fit a mean plus diurnal and semidiurnal components to the data, using the method of least squares. What we wish to discuss here is what happens in such an analysis when other types of time variation are present.

Let us assume that data are available at $t = t_i$, where

$$t_i = \frac{(i-1)}{n} 12 \text{ hrs},\tag{3}$$

$$i = 1, 2, 3, \ldots, n$$

and let

$$f^* = a_0 + a_1 \cos \pi t/12 \text{ hrs} + b_1 \sin \pi t/12 \text{ hrs} + a_2 \cos \pi t/6 \text{ hrs}$$
$$+ b_2 \sin \pi t/6 \text{ hrs}.\tag{4}$$

Let $f(t_i)$ be the actual observations (averages) at t_i. We wish to choose $a_0, a_1, b_1, a_2,$ b_2 to obtain a best least squares fit of $f^*(t_i)$ to $f(t_i)$. Let

$$I = \sum_{i=1}^{n} \left(f(t_i) - f^*(t_i) \right)^2.\tag{5}$$

The best least squares fit is obtained when

$$\partial I/\partial a_0 = \partial I/\partial a_1 = \partial I/\partial b_1 = \partial I/\partial a_2 = \partial I/\partial b_2 = 0.\tag{6}$$

Using Equation (5) we obtain

$$I = \sum_{i=1}^{n} f_i^2 + na_0^2 + a_1^2 C_{11} + b_1^2 S_{11} + a_2^2 C_{22}$$
$$+ b_2^2 S_{22} - 2a_0 A_0 - 2a_1 A_1 - 2b_1 B_1 - 2a_2 A_2$$
$$- 2b_2 B_2 + 2a_0 a_1 C_{10} + 2a_0 b_1 M_{10} + 2a_0 a_2 C_{02}$$
$$+ 2b_2 a_0 M_{20} + 2a_1 b_1 M_{11} + 2a_1 a_2 C_{12}$$
$$+ 2a_1 b_2 M_{21} + 2b_1 a_2 M_{12} + 2b_1 b_2 S_{12} + 2a_2 b_2 M_{22}\tag{7}$$

where

$$A_m = \sum_{i=1}^{n} f_i \cos \frac{m\pi t_i}{12}$$

$$B_m = \sum_{i=1}^{n} f_i \sin \frac{m\pi t_i}{12}$$

$$C_{jm} = \sum_{i=1}^{n} \cos \frac{j\pi t_i}{12} \cos \frac{m\pi t_i}{12} \tag{8}$$

$$S_{jm} = \sum_{i=1}^{n} \sin \frac{j\pi t_i}{12} \sin \frac{m\pi t_i}{12}$$

$$M_{jm} = \sum_{i=1}^{n} \sin \frac{j\pi t_i}{12} \cos \frac{m\pi t_i}{12}$$

and

$$f_i = f(t_i).$$

Equations (7) and (6) lead to five equations for a_0, a_1, b_1, a_2 and b_2:

$$na_0 + C_{10}a_1 + M_{10}b_1 + C_{02}a_2 + M_{20}b_2 = A_0 \tag{9}$$
$$C_{10}a_0 + C_{11}a_1 + M_{11}b_1 + C_{12}a_2 + M_{21}b_2 = A_1 \tag{10}$$
$$M_{10}a_0 + M_{11}a_1 + S_{11}b_1 + M_{12}a_2 + S_{12}b_2 = B_1 \tag{11}$$
$$C_{02}a_0 + C_{12}a_1 + M_{12}b_1 + C_{22}a_2 + M_{22}b_2 = A_2 \tag{12}$$
$$M_{20}a_0 + M_{21}a_1 + S_{12}b_1 + M_{22}a_2 + S_{22}b_2 = B_2. \tag{13}$$

These five equations are linear in a_0, a_1, b_1, a_2, and b_2, while A_0, A_1, B_1, A_2, and B_2 are linear in f. If f consists in components other than a mean plus diurnal and semidiurnal components, these other components will make 'spurious' contributions to a_0, a_1, b_1, a_2, and b_2. If we treat these other components separately, then the contribution of a combination of these components will equal the sum of the various contributions. We have investigated the following forms for f:

(a) $f(t_i) = \pi t_i / 12\text{hrs } t_i$ \hfill (14)

(b) $f(t_i) = +1, \quad i = 0, 2, \ldots$ \hfill (15)
        $f(t_i) = -1, \quad i = 1, 3, \ldots$

(c) $f(t_i) = \sin 2\pi t_i / 3 \text{ hrs } t_i$ \hfill (16)

(d) $f(t_i) = \cos 2\pi t_i / 3 \text{ hrs } t_i.$ \hfill (17)

We have also considered $n = 6$ and $n = 11$. Tables 2S.11 and 2S.12 show the contributions the above $f(t_i)$'s make to a_0, a_1, b_1, a_2 and b_2 for the two choices of n.

TABLE 2S.11

$n = 6$	fa	fb	fc	fd
a_0	1.31	0	8.83	5.10
a_1	−1.44	1.47	− 4.42	− .539
b_1	.386	−.395	−13.0	−8.05
a_2	.132	−.671	−4.45	−3.49
b_2	.228	−1.16	3.49	.425

TABLE 2S.12

$n = 11$				
	fa	fb	fc	fd
a_0	1.43	3.19	-1.03	.468
a_1	-1.50	$-.69$	.544	.586
b_1	.215	-4.80	1.54	$-.825$
a_2	.079	-1.77	.57	$-.464$
b_2	.27	.52	$-.452$	$-.487$

What is significant in Tables 2S.11 and 2S.12 is that spurious contributions to a_0, a_1, b_1, a_2 and b_2 may exceed the amplitudes of the various f's. Thus the condition that a field be largely due to steady, semidiurnal and diurnal components must be a very demanding condition indeed, if the calculated coefficients are not to be largely due to spurious contributions.

THE LUNAR ATMOSPHERIC TIDE AS REVEALED BY METEOROLOGICAL DATA

2L.1. Introduction

The lunar atmospheric tide, whose existence was recognized by Newton, was studied by Laplace, as indicated in Chapter 1, both theoretically, and from barometric observations. His attempt to determine it from eight years' observations, made four times daily at unequal intervals at the Paris observatory, was unsuccessful; he used only 4752 such readings, and after considering the probable error of the result he got, he decided that it was unreliable. Bouvard (1827) extended Laplace's calculations by including another four years' data, making twelve years in all, January 1, 1815 to January 1, 1827; from 8940 readings his result was an amplitude of 0.0176 mm, with maxima at 2h 8m and 14h 8m of lunar mean time. The great difference between this and Laplace's amplitude, 0.054 mm, confirmed Laplace's conclusion that the data were too few (in both cases) to provide a reliable result.

In 1843 Eisenlohr renewed the attempt with twenty-two years' additional data (1819–40), using all four daily readings. Unfortunately he departed from Laplace's excellent method of computation, which involved only *differences* between readings on the same day, thus eliminating the influence of the large changes of pressure from day to day (cf. Figure 1.1) characteristic of such latitudes as that of Paris. Eisenlohr rearranged his data according to the nearest lunar hour (0 to 23) at the time of each reading. With unlimited data this method would lead to a satisfactory result, and give the complete average change of the barometer according to lunar time, with the much larger solar daily change averaged out. But with his limited data the number of readings per lunar hour ranged from 1302 to 1377, and the hourly means were consequently differently affected by the great weather variations of pressure. The means showed a quite irregular variation from hour to hour. Thus his laborious effort, in many ways well-planned, was fruitless. He rightly concluded that his data were insufficient to determine L_2, and hence *a fortiori* were those of Laplace and Bouvard. He urged that *hourly* readings of the barometer be taken, so that in time, from a long series, L_2 might be determined.

This is what actually happened, long afterwards. In 1945, Daniel Kastler and Juliette Roquet, working in the Department of Mathematics at the Imperial College, London, determined the annual mean L_2 at Paris from 26 years' hourly data obtained at the meteorological and magnetic station in the Paris suburb of Parc St. Maur. The result was that the amplitude is 13.4μb (with probable error 2.7 μb); this is distinctly less than Bouvard's result (23.5 μb), and much less than that of Laplace (72 μb); the phase 105° implies maxima half an hour before upper and lower mean lunar transit. Rougerie (1957) extended this determination (see pp. 87, 89).

2L.2. The Tropical Lunar Air Tide

When Eisenlohr wrote (1843), L_2 had already been determined from tropical pressure data, though the result was not published until 1847, by Sabine. Around 1840, several British colonial observatories, magnetic and meteorological, were set up under Sabine's leadership. In 1842 the director (Lefroy) of the St. Helena Observatory successfully used his seventeen months' (August 1840 to December 1841) of bihourly weekday barometric eye readings to determine L_2. (No readings were taken on Sundays because of the Sabbatarian principles then supported in Britain.) His successor Smythe, with Sabine, confirmed the determination from hourly readings for three more years, October 1842 to September 1845. Sabine (1847) was even able to show from these data that the air tide was greater near perigee than near apogee.

In 1852 the director (Elliot) of the Singapore observatory determined L_2 there from five years' data, 1841 to 1845.

Later, when the Batavia observatory was established in 1866, these results stimulated its directors, Bergsma (1871, 1878) and Van der Stok (1882, 1885, 1887), to determine L_2 from their hourly barometric data, recorded photographically. Bergsma published the first results, for 1866 to 1868, in 1871. By 1905, L_2 at Batavia was really well determined, from 350000 observations covering forty years (ref. Figure 1.3).

2L.3. The Lunar Air Tide Outside the Tropics

Laplace (1799, 1827) stressed the need, in deriving results from observations, to determine the probability that their error lies within narrow known limits. This need has been overlooked or neglected by a multitude of those who, before and since his time, have vainly sought for lunar *monthly* meteorological variations. Eisenlohr, from 1833, was among these; but only a few of them have, like him, engaged in the more hopeful but still perilous search for lunar daily meteorological variations, in particular for L_2. Of these few, some, like Kreil in 1841, or later Bouquet de la Grye, used quite inadequate data – for one year only, or even for five years, like Neumayer (1867), who in 1867 failed to obtain consistent results for $L_2(p)$ from five years' hourly data (1858 to 1863) for Melbourne (lat. 38°S). Even Airy (1877), who used 180000 hourly values for Greenwich (51°N; for twenty years, 1854 to 1873), unwisely concluded that "we can assert positively that there is no trace of lunar tide in the atmosphere". Börnstein (1891) used only four years' data for Berlin and Vienna and Hamburg; for Keitum (55°N) he used ten years' hourly data (1878 to 1887), but found "no trace of a semidiurnal variation such as a lunar tide would produce"; he thought, however, he had found a definite lunar diurnal variation, a view shared also by Wegener (1915). Bartels (1927) showed that these conclusions completely misinterpreted the actual results that Börnstein had obtained, and that his supposed lunar diurnal variation was a purely chance effect, whereas L_2 was contained in his curves, though it was ill-determined. The misinterpretation sprang directly from the neglect of Laplace's advice to consider the probable accuracy of the results.

Morano (1899) used four years' data (1891 to 1894) for Rome (lat 42°N), although

Neumayer (1867) had failed, by the same method applied to five years' data for Melbourne, in a rather lower latitude, to obtain a reliable result. The validity of Morano's result, which was probably near the true value of L_2 at Rome, remained uncertain.

A new attempt (Chapman, 1918a) succeeded in determining L_2 from the Greenwich hourly data, by then available for sixty-four years. Two-thirds of the material was rejected; the only days used were those on which the barometric range did not exceed 0.1 in. This was the first certainly valid nontropical determination of L_2.

This investigation was planned and undertaken with guidance (Chapman, 1918b), in a very simple way, from the theory of random errors. According to this theory, if a single observation is subject to an accidental error e, the probable random error of the sum of N such (independent) observations is $e\sqrt{N}$, and that of their mean is $e/\sqrt{N}$. The moon produces a systematic (though very small) semidiurnal variation of the barometer; if this is combined from N lunar days' observations (each in the form of a sequence of lunar hourly values), by forming the sum of the values for each lunar hour, the sequence of sums will contain N times the lunar daily variation. It will, however, be affected also by the other causes of variation, particularly, outside the tropics, by the succession of cyclones and anticyclones. If these produce an average random departure e of any hourly value from the long-term barometric mean, they will contribute to each lunar hourly sum of N hourly values a random contribution of the order $e\sqrt{N}$. As N is increased, the regular lunar daily variation in the lunar hourly sequence of sums will increase proportionately to N, and the random contributions will increase, but proportionately only to $\sqrt{N}$. Thus, although e greatly exceeds the range of the lunar air tide at Greenwich, the systematic tidal effect will altogether overpower the random contribution, if N is taken large enough. In the sequence of lunar hourly *means*, L_2 is independent of N, whereas the random errors are of the order $e/\sqrt{N}$.

As the Greenwich data were used only for days of barometric range 0.1 in. or less, the average random departures e from each day's mean might be estimated as 0.01 inch. As N was 6457, $e/\sqrt{N}$ would be about 0.00012 inch.

Figure 2L.1 (full line) shows the mean lunar daily variation of Greenwich pressure obtained from these N days, by a method of rearrangement of solar hourly values according to lunar time. Happily, this method avoided a pitfall, then unsuspected but afterwards disclosed by Bartels (1927), associated with the use of selected barometrically 'quiet' days (Chapman, 1936).

The total range of pressure in Figure 2L.1 is less than 0.001 in., and the change from one lunar hour to the next averages about 0.00015 in. This exceeds the average random error in Figure 2L.1, namely, 0.00012 in., and the systematic nature of the lunar daily variation is clearly manifest. Apart from its meteorological and dynamical interest, this determination has great statistical interest as a remarkable illustration of the 'law of combination of random errors' – an example confirmed by many later air-tide determinations, most notably by that of the tidal variation of air temperature at Batavia (Chapman, 1932a, b).

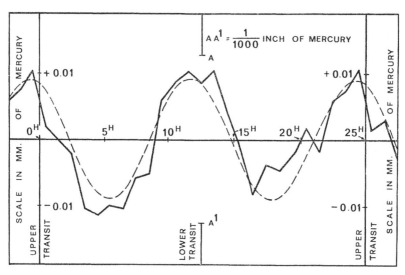

Fig. 2L.1.   Daygraph of $L_2(p)$ for Greenwich computed from 6457 quiet days' hourly barometric data, 1854–1917; the broken line shows the semidiurnal component of the daygraph, computed by harmonic analysis. After Chapman (1918a).

2L.4. The Month and the Lunar Day

The moon describes an orbit round the earth in a plane inclined at $5°.15$ to the ecliptic; the pole of the orbit revolves about that of the ecliptic once in 18.60 years, so that the inclination of the plane of the moon's orbit to the earth's equator varies between $23°.45 \pm 5°.15$, or $18°.30$ and $28°.60$. The moon's declination consequently changes during each passage round its orbit between maximum northern and southern values which may vary from $18\frac{1}{2}°$ (for instance, in the year 1941) to $28\frac{1}{2}°$ (in the year 1950); in midwinter, therefore, the full moon stood $10°$ higher at midnight in 1950 than in 1941. The mean distance of the moon from the earth is 384405 km, or 60.335 times the earth's radius (6371.2 km). The eccentricity of the orbit is considerable, and slightly variable; the mean ratio of the maximum distance, at apogee, to the minimum value, at perigee, is 1.1162, and the maximum ratio is 1.1411. The period from one apogee to the next is called the anomalistic month. The apogee revolves round the lunar orbit once in 8.8 years.

The moon's passage round the earth is accomplished in $27^d 7^h 43^m$ (the sidereal period), so that the mean lunar day, or average interval between two successive passages of the moon across any terrestrial meridian, is $24^h 50.47^m$, though the actual interval varies owing to the changing distance and orbit of the moon.

The moon revolves round the earth relative to the line OC through the sun's centre once in M days, where $M = 29.5306$. This period is called a lunation, or the synodic or lunar month. The moon's phases depend on the angle ν between the meridian half-planes through the sun and moon (reckoned positive when the moon is to the east of the sun). New moon corresponds to $\nu = 0$, and the values of ν at first eighth phase, first quarter (half-moon), full moon, and so on, are 45°, 90°, 180°, and so on. The phase is a measure of the moon's age, reckoned from new moon.

The east longitude τ of any station P relative to the meridian opposite to that containing the moon is a measure of the local apparent lunar time. Clearly, if t is the local apparent solar time (Figure 2L.2),

$$t = \tau + v, \tag{1}$$

if angular measure is used.

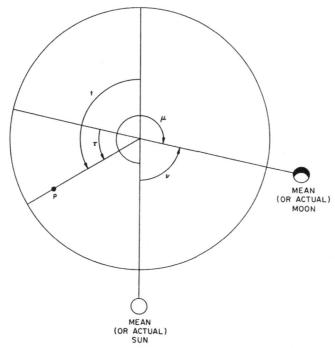

Fig. 2L.2.   To illustrate the relation between solar and lunar (apparent or mean) time and the lunar phase angle. After Sugiura and Fanselau (1966).

Mean instead of apparent solar time t is used (as in civil life) in calculating solar daily variations; because it proceeds uniformly, its use is more convenient in observatory work and records. The difference between apparent and mean time can be readily allowed for in any discussion concerning variations with apparent time (cf. Section 1.2A). By analogy with the mean solar time, a mean lunar time is introduced, determined by the motion of a fictitious mean moon, which is imagined to revolve uniformly round the earth. We shall now use t and τ to denote mean and not apparent solar and lunar time. Just as t is counted in 24 solar hours from midnight to midnight, τ is 0 for the lower transit of the mean moon and increases by 24 lunar hours up to the next lower transit. A mean lunar day equals 1.03505 solar days, or $24^{h}50.47^{m}$. If now we denote by v the angle between the meridian half planes containing the mean sun and mean moon, then (Figure 2L.2)

$$t = \tau + v ; \tag{2}$$

here t, v, and τ may be reckoned either in angular measure or in hours ($15° = 1$ hour). The age v of the mean moon completes a full cycle of 24 hours in the course of a synodic month; v increases by 0.81272 hours in the course of a mean solar day; correspondingly $v°$ increases by $12°.19080$ in each such day.

The differences between apparent and mean lunar time are much greater than those in the solar case. The following extreme cases illustrate the variability in the length of the apparent lunar day – the time interval from one transit to the next. In Berlin, on 1893 December 22–23, this interval was $25^h 8.6^m$ of mean solar time, but on 1913 August 18–19 it was only $24^h 38.7^m$, nearly half an hour less. The semi-diameter of the moon's disk was $16' 47.4''$ in the first case and $14' 43.0''$ in the second case; the actual moon's motion in its orbit was, in the first case, faster, because of its closer proximity to the earth, so that the apparent motion of the moon in the daily celestial rotation appeared slower. On the other hand, the length of the apparent solar day never deviates from that of the mean solar day by more than half a minute of time.

On 1930 January 13–15 the upper transits of the moon followed each other after an interval of $25^h 7.3^m$; the moon's semi-diameter had the high value $16' 47.4''$, as in the former example. About a quarter of a month before or after such a day on which the moon approaches the earth so closely, the greatest differences between apparent and mean lunar time are to be expected. Thus on 1930 January 9 the apparent moon culminated more than half an hour *before* the mean moon, but then speeded up in its orbital motion, and on January 21 culminated half an hour after the mean moon.

In spite of these irregularities, lunar geophysical reductions (other than those relative to the sea-tides) have, in the past, usually been based on the movement of the actual moon – largely because of the convenience of the data furnished day by day for the actual moon in the astronomical year-books. The tidal action of the moon, however, has been expressed, by strictly harmonic analysis (Section 2L.4A) of the tidal forces, in relation to the motion of the mean moon. Therefore it is preferable to use mean lunar time throughout.

Schmidt (1928), in his studies of the lunar daily geomagnetic variations at Potsdam, introduced the quantity given, for each day of Universal or Greenwich Time, by

$$\mu = 24 - v_u°/150° ,$$

where $v_u°$ signifies the phase angle of the mean moon at Greenwich noon on that day, reduced by such an integer multiple of $360°$ as to bring μ within the range from 0 to the unattained upper limit 24. Bartels and Fanselau (1937, 1938a) published a *Lunar Almanac* giving for the first day of every month, for the years 1850 to 1975 inclusive, the value of μ to two decimal places, together with other quantities related to the moon's mean apogee and perigee, and to its passage through the ascending node. Bartels and Fanselau (1938b) published *Moon Tables* giving the integer Mu, ranging from 0 to 23, nearest to the value of μ, for each day of the same period. This integer regresses from 23 to 0 in the course of each lunation.

However, it seems better to characterize each Greenwich day by the integer Nu from 0 to 23 nearest to the value of $v_u^\circ/15^\circ$ on that day, after modifying v_u° by an integer multiple of 360° so as to bring Nu within the said range; still better is to use the integer Nu' ranging from 0 to 11, equal to the integer nearest to $v_u^\circ/15^\circ$ after modifying v_u° by an integer multiple of 180° so as to bring Nu' within that range. Bartels (1954) gave daily values of Nu' (denoted by him by L) for the years 1902–52; Sugiura and Fanselau (1966) have published daily values of Nu' for the period 1850 to 2050. These tables are useful for computations to determine the influence of the main term (M_2) in the harmonic development (Doodson, 1922) of the lunar tidal potential. In computer programs to determine lunar geophysical effects from observational data, it may be more convenient to use the simple formula for v° (Section 2L.4A) to generate the number Nu' rather than to take it from the Tables.

2L.4A. THE MAIN HARMONIC COMPONENTS OF THE LUNAR TIDAL POTENTIAL

The lunar tidal potential V (cf. pp. 121–123) was analyzed harmonically (in the main) by Darwin (1901), who gave symbols for the principal terms. Doodson (1922) completed the analysis and gave a strictly harmonic development. His numerical constants have been improved as astronomical measurements of the quantities involved have gained in accuracy (Bartels and Horn, 1952; Bartels, 1957).

The main term, denoted (following Darwin) by M_2, for a point P in geographic latitude ϕ and at distance r from the center of the earth (whose mean radius we here denote by r_0) is

$$0.90812 (r/r_0)^2 G \cos^2 \phi \cos 2\tau,$$

where

$$G = 26206 \text{ cm}^2/\text{sec}^2.$$

The main term that depends on the varying distance of the moon from the earth is N_2, approximately lunar semidiurnal, given by

$$0.17387 (r/r_0)^2 G \cos^2 \phi \cos(2\tau - s + p),$$

The only remaining term that seems at all likely to have a perceptible effect on meteorological data (until the time series for them are very much longer than now) is the term O_1, approximately lunar diurnal, given by

$$0.37689 (r/r_0)^2 G \sin 2\phi \sin(\tau - s).$$

At the earth's surface the factor $(r/r_0)^2$ is unity. The horizontal component of the potential gradient seems the most likely to affect meteorological and ionospheric data. Its northward component X is $\partial V/r\partial\phi$, and its eastward component Y is $\partial V/r \cos\phi\partial\lambda$, where λ denotes the eastward longitude of the point on the earth. Thus the values of X derived from M_2 and N_2 vary with latitude on the earth as $\sin 2\phi$, and the the part derived from O_1 varies as $\cos 2\phi$; the former increase with latitude up to 45° and then decline again to zero; the O_1 (diurnally varying) part of X is strongest at the equator and poles, in opposite directions, with zero at 45° latitude.

The ϕ dependence of Y is the same as for the above terms of V; its semidiurnal part declines steadily from a maximum at the equator to zero at both poles; the diurnally varying part is zero at the poles and equator, with opposite maxima at $\pm 45°$ latitude.

In the above time factors,

$$\tau = t + h - s = t - v, \quad v = s - h, \quad 2\tau - 3 + p = 2t - 3v - h + p;$$

s denotes the longitude of the mean moon (s for *selene*, Greek for moon), h the longitude of the mean sun (h for *helios*, Greek for sun), and p denotes the longitude of the moon's perigee. The values of h, s and p are as follows:

$$s = 270°.43659 + 481267°.89057T + 0°.00198T^2$$
$$h = 279°.69668 + 36000°.76892T + 0°.00030T^2$$
$$p = 334°.32956 + 4069°.03403T - 0°.01032T^2 - 0°.00001T^3$$
$$v = s - h = -9°.26009 + 445267°.12165T + 0°.00168T^2$$

Here T signifies the time that has elapsed since Greenwich mean noon on 1899 December 31, measured in the Julian century (36 525 days) as unit; alternatively the middle term may be expressed in terms of t_d, the elapsed time in days; in that case the corresponding coefficients of T are respectively

(for s) 13°.176 396 730 246,   (for h) 0°.985 647 335 387,
(for p) 0°.1 114 040 802,   (for v) 12°.190 749 396 859,
(for $3v + h - p$) 37°.446 491 446.

The harmonic N_2 is the principal one involved in the lunar tidal effect of the changing distance of the moon from the earth. Its appearance in geophysical data has been verified by Bartels and Johnston (1940) in regard to the horizontal geomagnetic force at Huancayo, Peru, nearly on the magnetic dip equator. Just as the M_2 term in geophysical data is determined by dividing the days into twelve groups according to their Nu′ number, it is convenient in seeking an N_2 term to divide the days into twelve groups according to their N' integer defined as the one nearest to the Greenwich noon values of

$$12k + (3v + h - p)/30°;$$

here k is an integer chosen so that N' lies in the range 0 to 11; this grouping has the advantage that the Chapman-Miller method of analysis can be readily adapted to determine the N_2 term by substituting N' for Nu′ in the program, and changing d_{m2s} to 0.9323 for $S = 24$ or 0.9357 for $S = 12$. Thus the 12-fold grouping is preferable to the 8-fold grouping proposed by Bartels (1954), in a paper in which he gave the integers N nearest to the Greenwich noon values of

$$8k - (3v + h - p)/45°;$$

here k is an integer chosen so that N lies in the range 0 to 7. Whereas Nu' increases approximately twice by 12 in each lunation, N increases approximately three times by 12, and N decreases approximately three times by 8, in each lunation. (Bartels' table of N is not affected by errors in his account of these numbers; e.g., he wrongly stated that N decreases daily by 45°.3.)

The variation of the moon's distance from the earth is sufficient to make considerable changes in the intensity of the tidal force, which varies inversely as the cube of the distance; the variation about the mean value can range up to $\pm 20\%$. Consequently the apparent lunar and lunisolar daily geophysical variations produced by the lunar tide increase and decrease throughout the anomalistic month. The determination of N_2 in geophysical data is preferable, as a way of studying the influence of the changes of lunar distance, to the method occasionally used in the past, of determining L_2 for groups of days centered on or lying between the epochs of apogee and perigee of the actual moon (the combination of M_2 and N_2 produces 'beats' in the apparent values of L_2).

2L.5. Methods of Computation of L from Observed Data; Early Methods based on Apparent Lunar Time

Many different methods have been used to determine lunar daily variations in geophysical data. The method used by Lefroy, who made the first successful determination of $L_2(p)$, for St. Helena (15°57'S), was described by him as follows, in a report to Sabine dated June 1, 1842:

The corrected height of the barometer (i.e., the reading reduced to 32° Fahrenheit) at the hour of observation nearest the moon's meridian passage for every day has been entered in a central column; and in parallel columns headed -2^h, -4^h, &c and $+2^h$, $+4^h$, &c have been entered the '(bihourly)' observations at 2^h, 4^h, &c before and after the central observation. A mean has been taken for the observations included in each lunar month. It appears from the seventeen months thus examined, that a maximum of pressure corresponds to the moon's passage over both the inferior and superior meridians, being slightly greater in the latter case; and that a minimum corresponds nearly to the rising and setting, or to $\pm 6^h$. The average of the seventeen months gives the respective pressures as follows, viz. –

	inches
Moon on the meridian	28.2714
Moon on the horizon	28.2675

The difference being .0039 inch.

From October 1, 1842 the readings of the barometer at St. Helena were made hourly. Smythe, Lefroy's successor, somewhat refined the above method; he subtracted the mean solar daily variation derived from each lunar month, from the hourly observations, before re-ordering them in rows of 12 values preceding and 12 following the value for the hour nearest to the moon's central meridian passage each day. Sabine continued this procedure with further data from St. Helena, but used the mean $S(p)$ for the calendar months, instead of the rather more exact course of treating the lunar months as Smythe had done. Smythe and Sabine both gave lunar daily sequences of 25 mean values, for whole or half years. They did not make a harmonic analysis of their sequences, or consider the probable error. The result obtained by harmonic analysis of their combined sequence was later given by Chapman and Westfold (1956).

Smythe remarked that sometimes there were only 23 (solar hourly) readings between successive meridian passages of the moon; in that case he entered the reading midway between the passages, in both the columns -12 and $+12$.

The Sunday intermission of the hourly readings somewhat reduced the number of usable days' data. The introduction of pen or photographic recording finally removed this difficulty affecting the British stations.

The sequences of 25 equidistant readings obtained by the above methods were sometimes graphed and readings taken from the graph at 24 equal intervals, for ease of harmonic analysis; alternatively the formulae for harmonic analysis of 25 equidistant values were used.

At some stations, after the introduction of automatic recording, the records were measured not only at the solar hours but also re-measured at lunar hours, but the varying length of the intervals between meridian passages of the moon complicated this laborious procedure.

In the determination of $L_2(p)$ at Greenwich (Chapman 1918a), based on the readings taken from the photographic record for the 64 years 1854–1917, the data used were confined to 6457 days, about a third of the whole – the selected days being those on which the range did not exceed 0.1 inch. The method of treatment was unusual; in each row of solar hourly readings a red vertical mark was made preceding the entry for the hour next before the moon's transit on that day (or before the hour of transit when this coincided with a solar hour). To quote from the account of the further procedure:

In transferring the entries for the quiet days from the original sheets to others arranged according to lunar time, days from each calendar month were kept on separate sheets; on each sheet the rows were entered in chronological order, the year being indicated in the first column.

On the 'lunar' sheets there were 25 hourly columns following upon the one just mentioned. These columns were headed $0\frac{1}{2}$ h, $1\frac{1}{2}$ h, and so on. In these were copied the hourly entries for each quiet day, beginning with the one immediately to the right of the red mark indicating the time of lunar transit. On the average this reading would be one-half solar hour after the latter epoch, regarded as 0 h of the lunar day. The lunar day contains approximately 25 solar hours, so that the last entry in any quiet-day row (at 23 h. solar time) would correspond to a lunar time one hour earlier than that for the entry immediately above it on the original sheets (at 23 h of the preceding solar day). Hence, when the entries to the right of the red mark on any quiet-day row had been copied on to the appropriate lunar sheet, the next entry copied was the one just referred to, and then followed those for 0 h, 1 h, and so on, for the earlier part of the quiet day, until the red mark was again reached from the left. Thus the 25 entries in any lunar row corresponded to a lunar day of 25 hours, commencing at 23 h on the day previous to the quiet day itself; this lunar day was, however, broken into two pieces, so as to make the part *before* the lunar transit follow that beginning with the hour immediately *after* lunar transit. The discontinuity thus introduced occurred, of course, at all lunar hours, and it was assumed that its effect would average out with the other irregularities present.

The regular solar daily barometric variation was not abstracted from the observations before transference to the lunar sheets. The transference, however, was not simply a process of copying. Previous experience of such work as this suggested that little or nothing would be gained by copying the hourly entries to the third decimal place (representing units of 0.001 inch). The earlier decimals vary so much from day to day, on account of the large irregular fluctuations in the barometer, that no real advantage accrues from the retention of the third decimal, provided the second decimal figure is raised by one unit when the third figure exceeds 5. When the third figure was 5, the next *even* figure was adopted for the second decimal; thus 29.875 would be read as 29.88, and 29.865 as 29.86.

Further, since the range on the selected days did not exceed 0.10, it was convenient to subtract from all the entries for any one lunar row such a number, consisting of so many inches and whole tenths of an inch, as would render the least and greatest remainders less than ten and twenty hundredths respectively. Thus, if the greatest and least of the 25 entries on the original sheets were, in any particular case, equal to 29.88 and 29.79 respectively (after the third decimals had been allowed for mentally), 29.70 would be subtracted throughout that lunar row, and the greatest and least entries on the lunar sheet would be 18 and 9. No decimal points were inserted on copying, so that the unit on the lunar sheets represented 0.01 inch.

After the copying had been done, on this plan, the sums were taken for each column. The results for each of periods 1854–73, 1874–93, and 1894–1917 were kept separate, so that it was possible to form the complete sums for the following three threefold subdivisions of the material: (a) all days, of whatever month or distance group, in each of the above three periods; (b) days, drawn from the whole period 1854–1917, grouped according to the three seasons, summer (May, June, July, August), equinox (March, April, September, October), and winter (November, December, January, February); and (c) days, likewise drawn from the whole period, grouped according to the moon's distance, as indicated by the numbers 14, 15, 16 representing its semi-diameter. Nine 25-hourly lunar diurnal inequalities were thus obtained, being the sums of the lunar entries for about 2000 days in each case. These nine inequalities, and a tenth which was derived from all the 6457 days, were subjected to harmonic analysis, to determine the diurnal and semidiurnal component variations. A correction was duly made to allow for the fact that the lunar entries were at the half-hours, $0\frac{1}{2}$ h, and so on, and not at lunar hours. The Fourier coefficients were then divided by the number of days to which corresponded the sequence of sums analysed, and the results were, finally, expressed in units of 0.0001 mm of mercury.

The semidiurnal harmonic coefficients for the three groups of years agreed fairly well with each other, though the diurnal coefficients were plainly accidental. But in other cases there were differences between the values obtained for $L_2(p)$, which were interpreted as due to a residue of $S(p)$ not fully eliminated by the averaging. Hence a method was devised and applied that removed this influence, and this brought reasonable agreement between the various determinations. Later Bartels (1927) estimated the probable error of the three seasonal values and the annual mean of $L_2(p)$.

In the same paper Bartels showed that there is a danger in using only quiet days' data in determining $L_2(p)$. This is because, as he found, such days in the main occur near the epochs of maximum barometric pressure, when the general shape of the barograph is convex upward. In the ordinary method of rearranging the solar hourly values according to lunar time, this convexity is not eliminated; thus it remains in the resulting 'lunar' daily sequences, added to the true $L_2(p)$. Its range is comparable with or at some stations greater than that of $L_2(p)$, in temperate latitudes. Thus the harmonic coefficients do not represent only $L_2(p)$, and may be completely misleading. Chapman did not know of this when using the 6457 Greenwich days as the basis of his determination; but he unconsciously avoided the danger because of the peculiar plan he adopted, of putting the later part of each solar day before the earlier. In seeking to determine $L_2(p)$ from a long series of hourly data from Glasgow, Robb and Tannahill (1935) incurred the pitfall; the convexity variation in their mean lunar sequences exceeded L_2; Chapman (1936) attempted to correct the results by allowing for the convexity, but this was too large to enable a reliable correction to be made. Earlier Robb had used the method adopted for the Greenwich data, for many years of Glasgow data, and his results thus obtained showed that $L_2(p)$ at Glasgow is small for the latitude. Slightly earlier, Chapman and Austin (1934) had discussed $L_2(p)$ at Buenos Aires, using mainly almost all days, but also, in part of their work, selected quiet days. The results from the latter showed the convexity effect, and a correction was made for it, assuming that the convexity was parabolic; but at Buenos Aires, as at Glasgow, $L_2(p)$ is small for the latitude, and the results from almost all days were more reliable than those for quiet days, because of the uncertainty of the correction for the convexity of the barograph on those days. It would have been possible to obtain a better result from the quiet day data had the solar daily variation characteristic of these days been removed in the analysis, instead of only the monthly mean solar daily variation.

Variants of the methods just described, of determining L using solar hourly values and apparent lunar time, have been devised by several authors. Egedal (1956) has described 'two simple methods' (1956), one of which he applied to eleven years' record of magnetic declination at Rude Skov; the other he applied to a further 33 years' of the same record. Unlike $L(p)$, the L for the magnetic elements contains significant terms that are not semidiurnal, as was first discovered by Broun (1874); but Egedal, like many others before and since, did not make the division of his data into different groups of days at successive lunar phases, necessary to determine the non-semidiurnal terms.

In his first method the solar daily variation is not removed from the hourly data before these are re-arranged according to lunar time; the hourly data are simply re-written in rows of 24 values, taking for each lunar hour (1/24 of a lunar day) the hourly value most nearly coinciding with that time. This means that on most days one hourly value is superfluous; to use also this value to some extent, for the lunar hour nearest to the full solar hour the mean of the preceding and succeeding hourly values is used (the same method had been applied by Chambers in 1887 and Moos in 1910 to Bombay magnetic data).

In the other method, the solar hourly values are re-arranged in 24 rows of 28, 29 or occasionally 30 values each (Rooney, 1938). Egedal's description of this method is as follows:

As the treatment is similar for hourly and bihourly values only the procedure used for hourly values will be described. First the hourly mean values for the interval 0^h-1^h GMT have been treated, next the values for the interval 1^h-2^h and so on. The values for a considered hour are written down according to the age of the moon starting in the first row with the value of the first day for which the lunar age at 0^h in Greenwich is nearest to 0 and writing the following value in the second row and so on to the 30th row (lunar age 29). Hereafter the following values are placed in the second column and so on. When all values for the considered year are placed on the list the departures of the mean values for each lunar age from the common mean are computed and corrections for the secular variation are made, considering the secular variation to vary proportionately with time through the synodic month (29.53 days). In this way 24 sets of departures are obtained, each set consisting of 30 values.

Again in this method there is no removal of S from the data. For further details of the later stage of the computations using this method of retabulation of the data, the reader is referred to the original paper by Egedal.

The use of apparent lunar time in these methods is not essential to them, but was adopted only because of the ease with which the times of actual lunar transit can be obtained from the Nautical Almanac or other ephemeris. But it involved the irregularities of the length of the apparent lunar day. In geomagnetic lunar studies, Schmidt (1928) was a pioneer in following the practice adopted in modern sea tidal computations (Doodson, 1922), of using mean lunar time, and making the analysis truly harmonic. Chapman and his collaborators made many determinations of L in meteorological and geomagnetic data from 1914 onwards, using apparent lunar time as the basis, until about 1938. Then he changed to the use of mean lunar time, at the suggestion of Doodson, and in connection with a new method devised jointly with Miller (Chapman and Miller, 1940) (here referred to as the C-M method). It is the one now generally adopted (with slight variations) in geophysical studies of L; the older 'apparent lunar time' methods may be considered obsolete.

2L.6. The Chapman-Miller (or C-M) Method for Meteorological Variables

This is a simplified version of a method proposed by Chapman (1930), that has apparently never been applied. It was devised to improve the determination of L in cases where this is overlaid not only by S, but also by large irregular fluctuations – for example, to determine $L_2(p)$ for stations in high latitudes, where weather produces large barometric variations. The C-M method cuts down the eliminatory procedure proposed by Chapman to the allowance merely for the non-cyclic variation, either for each day, or more usually, in meteorological studies, in the mean for sets of days grouped according to the lunar phase number Nu'. Here the method is outlined only for analyses in which, as for meteorological variables, L_2 is the significant lunar harmonic to be determined; its application to geomagnetic and ionospheric data, whose variations usually include also non-semidiurnal (lunisolar) components, is more complicated.

The meteorological data considered are taken to be instantaneous values (as they usually are, for example, for air pressure and temperature), not hourly means as in the case of modern geomagnetic hourly values.

For all these kinds of data, the basic material for the determination of L consists of S values per day at equal time intervals, namely hourly ($S=24$) or bihourly (even or odd hours: $S=12$), or even trihourly. To eliminate the non-cyclic variation we take sequences of $S+1$ terms. The hours to which the first and last refer may, for example, be either 0 and 24 or 1 and 25.

In any section of the material, for example, for all the months January, or all the seasons j, over a period of years, the days are divided into 12 groups according to their lunar phase integers Nu'. For each group, sums are taken of the daily sequences of values; thus we have 12 sum-sequences, one for each value of Nu' (note that in formulae, v' will be written to signify the integer Nu'; v' is to be carefully distinguished from v; the latter is an angle, $0°$ to $360°$, whereas v' or Nu' is an integer ranging from 0 to 11).

The subsequent procedure involves two processes of harmonic analysis, called primary and secondary. The first, primary, harmonic analysis is applied to the 12 sum-sequences, allowing for the non-cyclic variation. In meteorological studies only the semidiurnal component S_2 need be determined; in geomagnetic and ionospheric studies the first four harmonics are calculated. Thus we obtain $A_2(v')$ and $B_2(v')$, the coefficients of S_2 for each sum-sequence v' ($=$Nu'), formed from $N(v')$ days. If there were no lunar daily variation, all the A's and all the B's, after division by their corresponding $N(v')$, would be the same, except for accidental error. But because of the presence of L_2, given by

$$L_2 = l_2 \sin(2\tau + \lambda_2) = l_2 \sin(2t + \lambda_2 - 2v), \tag{1}$$

$S_2(v')$ will be given by

$$S_2(v')/N(v') = S_2 + l_2' \sin(2t + \lambda_2 - 30° \, v'). \tag{2}$$

Here the first term is the true solar semidiurnal variation. The second term is the averaged contribution made by L_2 to the calculated semidiurnal variation on the solar days of phase integer v'. In (1), v represents a lunar phase angle that increases by $360°$ from one new moon (phase $0°$) to the next; on any solar day of the v' set its mean value will lie between $15°\, v' \pm 7.5°$, or between these values plus $180°$; in (2) the averaged value of $-2v$ will be $-30°\, v'$. Owing to this phase spread, and because the harmonic analysis refers to a solar, not lunar day, l_2' in (2) is slightly less than l_2.

On a harmonic dial, $S_2(v')$ is represented by the sum of the dial vector corresponding to S_2, with amplitude s_2 and phase σ_2, and a vector for the second term in (2), of amplitude l_2' and phase $\lambda_2 - 30°\, v'$. Thus as v' increases from 0 to 11, the second vector regresses in phase by $330°$, turning in the clockwise direction. The values of $A_2(v')$ and $B_2(v')$ are given by:

$$A_2(v')/N(v') = A_2 + l_2' \sin(\lambda_2 - 30°\, v'), \tag{3}$$

$$B_2(v')/N(v') = B_2 + l_2' \cos(\lambda_2 - 30°\, v'). \tag{4}$$

Each set of 12 values of $A_2(v')$ and of $B_2(v')$, and also the set of 12 numbers $N(v')$, is harmonically analyzed (the secondary analysis), as follows: the sums are all from $v'=0$ to $v'=11$;

$$A_{p,A} = \sum A_p(v')\cos(30°\, v'), \quad A_{p,B} = \sum A_p(v')\sin(30°\, v'), \tag{5}$$

$$B_{p,A} = \sum B_p(v')\cos(30°\, v'), \quad B_{p,B} = \sum B_p(v')\sin(30°\, v') \tag{6}$$

$$N_{1,A} = \sum N(v')\cos(30°\, v'), \quad N_{1,B} = \sum N(v')\sin(30°\, v'). \tag{7}$$

In the case of meteorological data $p=2$, as above indicated; in geomagnetic and ionospheric studies, other values of p, namely 1, 3, 4, are also taken into account.

Thus in the meteorological case we obtain from the secondary harmonic analysis the 6 numbers $A_{2,A}$, $A_{2,B}$, $B_{2,A}$, $B_{2,B}$, $N_{1,A}$, $N_{1,B}$. From these we calculate U_2, V_2 as follows (here taking $p=2$):

$$U_p = (A_{p,A} + B_{p,B}) - (A_{p,N} N_{1,A} + B_{p,N} N_{1,B})/N \tag{8}$$
$$V_p = (B_{p,A} - A_{p,B}) - (B_{p,N} N_{1,A} - A_{p,N} N_{1,B})/N \tag{9}$$

Here

$$N = \sum N(v'), \quad A_{p,N} = \sum A_p(v') \quad B_{p,N} = \sum B_p(v'). \tag{10}$$

The next step is to calculate l_2 and λ_2' from the equations:

$$l_2 \sin \lambda_2' = U_2/Kd_{mpS}, \quad l_2 \cos \lambda_2' = V_2/Kd_{mpS}. \tag{11}$$

Here

$$K = (e_R SN/2)(1 - N_1^2/N^2), \quad N_1^2 = N_{1,A}^2 + N_{1,B}^2 \tag{12}$$
$$e_R = (\sin \theta_R)/\theta_R, \quad \theta_R = \pi/R \quad R = 12; \quad 1/e_R = 1.01152 \tag{13}$$
$$d_{mpR} = \frac{\sin \pi(m-p)}{S} \left\{ \cot \frac{\pi(m-p)}{S} + \cot \frac{\pi p}{S} \right\} \tag{14}$$

and $m = 2(1 - 1/M)$, $M = 29.5306$, the number of mean solar days in a lunation. For $S = 24$ and $S = 12$ the values of d_{m2S} are as follows:

	$S = 24$	$S = 12$
d_{m2R}	0.9596	0.9619.

Finally a phase correction must be applied to λ_2' to obtain λ_2. The correction (to be added to λ_2') is given in degrees by

$$\lambda_2 - \lambda_2' = 2L'/M - 15°mH' + m(L - L')°. \tag{15}$$

Here L denotes the longitude of the station whose data are under discussion, measured from Greenwich (in degrees, reckoned positive if westward, negative if eastward, up to 180°); L' is the longitude (similarly measured and reckoned) of the meridian of time reckoning with respect to which the data are tabulated; and H' is the solar hour of the initial value of each daily sequence, according to the same time-reckoning.

Modern geomagnetic hourly data are not instantaneous values, like barometric readings, but means over hourly intervals; and some meteorological data, such as wind and rainfall, may likewise be mean values over an hour. In this case, the lunar amplitude l_2 (and its probable error) needs to be enhanced by the factor $1/e_R$ defined by (13); its value is 1.01152.

2L.6A. USE OF THE INTEGERS Mu (OR μ) INSTEAD OF THE INTEGERS Nu OR Nu' (OR v')

Until 1954, when Bartels (1954) gave a table of values of Nu' (or v') (there denoted by L), several lunar analyses of meteorological and geomagnetic data had been made using Schmidt's lunar phase integers Mu (μ) as the basis; and Chapman and Miller (1940) expounded their method in relation to these numbers. Tschu (1949) described the practical application of the method on this basis; Chapman (1952) corrected his account of the calculation of the probable errors involved. The choice of μ or v' as the basis is immaterial up to the end of the secondary harmonic analysis. But if μ is used, the signs of $N_{1,B}$, $A_{p,B}$ and $B_{p,B}$ in the Equations (8), (9) are to be reversed.

2L.6B. THE COMPONENTS S_p

In the course of the C-M method of determining L, the analysis gives also S_2 (and the other components S_p if these are calculated in the primary harmonic analysis, for values of p other than 2). Thus let

$$A'_p = \frac{1}{12} \sum \frac{A_n(v')}{N(v')}, \qquad B'_p = \frac{1}{12} \sum \frac{B_p(v')}{N(v')},$$

$$s'^2_p = A'^2_p + B'^2_p, \qquad A'_p = s'_p \sin \sigma'_p, \qquad B_p' = s'_p \cos \sigma'_p,$$

then

$$s_p = s'_p/6e_R, \qquad R = 24/p, \qquad \sigma_p = \sigma'_p - 15° \, pH' + p(L - L')°.$$

2L.7. Vector Probable Errors

In all cases of physical measurement it is desirable to give some estimate of the accuracy of the result. This is particularly true for lunar geophysical terms in meteorological data, because they are usually of very small amplitude, and overlaid by much larger solar daily changes, and also at many places by large irregular variations associated with weather changes.

Figure 1.3 shows a harmonic dial for $L_2(p)$ for Batavia, with $n(=40)$ points, all of equal weight, because each indicates the determination for one calendar year. Assuming that their plane distribution around their centroid C is Gaussian and symmetrical, the probable error r_0 of the 40-year mean determination given by C is $r/(n-1)^{1/2}$, where r denotes the probable error for each of the yearly points; this is given by $r=0.989d$, where d denotes their mean distance from C. If the true value of the 40-year mean $L_2(p)$ were known, r_0 would be $r/n^{1/2}$. But the denominator is changed to $(n-1)^{1/2}$ because C itself is determined from the n dial points, leaving only $n-1$ independent quantities, or, as the statisticians say, $n-1$ degrees of freedom (James and James, 1959, p. 163).

When L is determined by the C-M method, instead of finding its probable error from several determinations of L, it may be found for each single determination, from the 12 dial points for the results of the primary harmonic analysis. They are given by the coordinates $A_2(v')/N(v')$, $B_2(v')/N(v')$; these are represented by (3), (4) of Section 2.L.6 as the ends of a vector with components A_2, B_2 representing S_2, and a contribution from L_2, varying in phase from group to group, that is, depending on Nu'. But there is also a contribution of accidental character from the irregular variations. Were this not so, the 12 dial points would lie on a regular 12-sided polygon, but the accidental contributions distort the polygon. Let

$$\Delta A_2(v') = A_2(v') - N(v') A_2/N,$$
$$\Delta B_2(v') = B_2(v') - N(v') B_2/N.$$

From these the components of the error vectors are derived by the following equations:

$$\Delta'_A(v') = \Delta_A(v') \cos 30° v' + \Delta_B(v') \sin 30° v' - N(v') U_2/N$$
$$\Delta'_B(v') = \Delta_B(v') \cos 30° v' - \Delta_A(v') \sin 30° v' - N(v') V_2/N.$$

The root mean square of the amplitudes of the error vectors is given by

$$E = (1/12) \sum_{v'=0}^{11} \{\Delta' \ (v')^2 + \Delta' \ (v')^2\}^{1/2}.$$

The probable error r of any one dial point is given by

$$r = 0.9394E.$$

The probable error r_S of S_2 is $r/(n-2)^{1/2}$ where $n=12$, because we have determined the amplitudes of both S_2 and L_2 and r from the 12 points, losing two degrees of freedom. The probable error of r_L of L_2 is $1.01152 r_S/d_{mpS}$; the first additional factor corrects for the spread of $v/15°$ around the integers Nu'. Clearly r_S and r_L are nearly the same, and r_S is much smaller relative to s_2 than r_L is to l_2. It is desirable always to quote r_L in connection with any determination of L_2, but it does not seem necessary to quote also r_S in connection with the associated determination of S_2.

2L.8. The Determination of L_2 from Only a Few Meteorological Readings per Day

Bartels (1938) determined $L_2(p)$ for the African coast town Dar es Salaam from baro-metric readings made only three times daily, at 7, 14 and 21 hours, local mean solar time, over a period of 17 years. Like Laplace he dealt with *differences* between readings on the same day. He eliminated from his calculations the changes of mean level from day to day, and the non-cyclic variation (assumed to proceed uniformly) in the course of each day. He grouped the differences according to Schmidt's re-gressing phase integers μ, and determined the semi-monthly variation in the twelve mean differences. From these he was able to infer the values of l_2 and λ_2. Haurwitz and Cowley (1967) have applied his method, using the integers Nu' instead of μ. Their explanation of the method is followed here, in terms of Nu'.

Let h', h, h'' denote the local mean solar times (in hour units) at which the readings are taken; let $h' < h < h''$. Reckon local mean solar time t from midnight ($h=0$) on day zero. Numbering the days as $0, 1, 2, ..., r, ...$, then at hour h of day r, the time t, reckoned in angle at the rate $360°$ per mean solar day, is given by

$$t = 360°r + 15°h. \tag{1}$$

Let v_0 denote the mean moon's phase angle at time $t=0$, and let v, v_r denote its values on day r at hour h and at noon ($h=12$). Because v progresses at $1/M$ of the rate of progress of t, where $M=29.5306$ (see Section 2.L4), we have

$$v = v_0 + (360°r + 15°h)/M, \quad vr = v_0 + (360°r + 180°)/M, \tag{2}$$

Hence

$$v = v_r + 15°(h - 12)/M. \tag{3}$$

At any time t the pressure p is expressible as

$$p = p_r + A + S + L.$$

here p_r denotes the mean value of p on the day considered, A is an 'accidental' part, and S, L denote the contributions from the daily variations, solar and lunar. We suppose that in the averaging processes to be described, the influence of A and S on the results is reduced to insignificance compared even with the small part of p corresponding to L_2. Consequently henceforward in this explanation only L is mentioned. It is expressible thus (omitting the subscripts (2) of l and λ):

$$l \sin(2\tau + \lambda) \quad \text{or} \quad l \sin(2t = 2v + \lambda). \tag{4}$$

Thus at hour h on day r, the L_2 contribution to p is given by

$$L(r, h) = l \sin(720°r + 30°h - 2v + \lambda) \tag{5}$$
$$= l \sin\{30°h - 2v_r - 30°(h - 12)/M + \lambda\} \tag{6}$$
$$= l \sin(30°hg - 2v_r + \lambda + 360°/M) \tag{7}$$

where

$$g = 1 = 1/M = 0.966137. \tag{8}$$

It is clearly allowable to add to v_r in (8) whatever integer multiple of $\pm 180°$ will bring it into the range 0 to 180°; let us suppose this is done. With each day r there is associated a lunar phase integer v', in the range from 0 to 11, given by

$$v' = v_r/15° + \delta, \quad v_r = 15° v' + \sigma \tag{9}$$

where

$$-1/2 \leqslant \delta < 1/2, \quad -7°.5 \leqslant \sigma < 7°.5. \tag{10}$$

Here we take v to refer to Greenwich noon on the day considered. Let the average value of $L(r, h)$ taken over all the days associated with the phase integer v' be denoted by $L(v', h)$; then if there are many days in the group, we have very approximately

$$L(v', h) = f l \sin(2v' - 30°hg - \lambda - 360°/M + 180°), \tag{11}$$

where

$$f = \sin(\pi/12)/(\pi/12) = 1/1.01152. \tag{12}$$

By harmonic analysis of the sequence of 12 mean values $L(v', h)$ for $v' = 0, 1, 2, ..., 11$, we obtain the amplitude $f l$ and the phase angle $180° - 360°/M - \lambda - 30°hg$, and hence the values of l and λ can be found. But as in Eisenlohr's analysis of the Paris barometric data, the day to day changes of barometric level enter into the averaging, and the data may be too few, as in his case, to reduce their average effect on the deduced harmonic component of the $L(v', h)$ sequence to an adequately small amount.

Thus instead we may analyze the sequence of differences $L(r, h') - L(r, h)$ grouped according to v' and averaged; this eliminates the influence of the day to day changes of level, but not the non-cyclic variation during this interval of the day considered. Clearly

$$L(r, h') - L(r, h) = 2l \sin\{15° g(h' - h\} \cos\{15° g(h + h') -$$
$$2v_r + \lambda + 360°/M\}$$

and

$$L(v', h') - L(v', h) = l' \sin(2v' + \lambda'), \tag{13}$$

where

$$l' = 2fl \sin\{15° g(h' - h)\}, \quad \lambda' = -\lambda - 15° g(h + h') + 90° - 360°/M. \tag{14}$$

Thus if we analyze the sequence of mean differences (13) and so determine l' and λ', we can infer the values of l and λ from (14).

If $15° g(h' - h)$ were $\pm 180°$ the method would fail, because then $l' = 0$; if this angle is $90°$, the time difference $h' - h$ gives the largest value of l', namely $2fl$. If $h - h' = 7$, as at Dar es Salaam, l' is almost the same, being less by the factor 0.969.

When there are three readings daily, as supposed above, with $h - h'$ and $h'' - h$ both not very different from 6 or 7, it is convenient to deal with group averages of $L(r, h') - L(r, h) + L(r, h'') - L(r, h)$. The harmonic analysis leads to the following values:

$$l' \sin(2v' + \lambda') + l'' \sin(2v' + \lambda''),$$

where l'' and λ'' are given by (14) with h'' substituted for h'. If $h - h' = h'' - h$, as for the Dar es Salaam data, this procedure eliminates the non-cyclic variation; if not, the elimination can be made, using the value of this variation (for the day) given by the difference between the values at hour h' on the given day and the next day. This considerably complicates the formulae, and is not considered here. When the two intervals $h - h'$ and $h'' - h$ are equal, the formulae are as follows:

$$l = \{(l' \cos \lambda' + l'' \cos \lambda'')^2 + (l' \sin \lambda' + l'' \sin \lambda'')^2\}^{1/2}/4f \sin^2 15° (h' - h)$$
$$\lambda = -\lambda' - 360°/M - 7°.5 \, g(h' + 2h + h'').$$

To find the probable error of the result, it suffices to divide the data into m equal parts, and to obtain the L_2 dial point for each, as above. The centroid of the m points C gives the overall mean L_2, and its probable error is $0.9394 \, d/(m-1)^{1/2}$, where d denotes the mean distance of the m dial points from C.

2L.9. The Lunar Semidiurnal Barometric Tide $L_2(p)$

The many unsuccessful attempts to determine $L_2(p)$ included those of Laplace, Bouvard and Eisenlohr for Paris (Sections 1.1, 2 L.1), of Airy for Greenwich, of Neumayer for Melbourne, and of Börnstein for Berlin and Vienna (Section 2L.3); the main cause of failure was inadequate material. The first successful meteorological determinations of L were those given in Table 2L.1.

During the period covered in this table many determinations had been made also of the lunar *geomagnetic* tide (Chapman and Bartels, 1940, Chap. 8). Chapman's interest in lunar influences in geophysics began with studies in this field (1913, 1919a), which then led him to consider the simpler causative atmospheric tide determinable from the barometric records. Bartels' progress was in the opposite direction. For more than three decades they stimulated also other workers (called by Bartels the

TABLE 2L.1

Early determinations of $L_2(p)$. The unit for the amplitudes and probable errors is 1 microbar (μb)

Stations (s)	Years' data	Author(s)	Publication
St. Helena	1842–45	Lefroy, Sabine, Smythe	Sabine (1847)
Singapore	1841–45	Elliot	Elliot (1852)
Batavia	1866–1905	Bergsma, Van der Stok	Observatory Yearbooks
Rome	1891–94	Morano	1899
Greenwich	1854–1917	Chapman	1918a
Batavia	1866–95	Chapman	1919b
Hongkong	1885–1912	Chapman	1919b
Aberdeen	1869–1919	Chapman, Falshaw	1922
Mauritius	1867–1915	Chapman	1924a
Tiflis	1880–1905	Chapman	1924a
Potsdam	1893–1922	Bartels	1927
Hamburg	1884–1920	Bartels	1927
Keitum	1878–87	Bartels	1927

Society of Lunatics!) who made or studied determinations of L from meteorological or magnetic records. They included Miller, Westfold and Tschu in England, Fanselau, Schneider, Kertz and Siebert in Germany, and Rougerie in France.

In the studies by Chapman and his collaborators up to 1930, the data were rearranged on handwritten sheets; they took apparent lunar time as basis; at most a desk calculator was used. Then these primitive methods, differing little from those first used by Lefroy in 1842, gave place in 1930 to the use of punched cards (45 columns), and of sorting and tabulating machines (freely loaned and serviced by the British Tabulating Machine Company), with the aid and guidance of L. J. Comrie. (Bartels, however, always used handwriting methods, applied to hour-to-hour differences, allowing simple checks on the copying and subtractions; he also pioneered in the harmonic analysis of each day's data.)

A small computing staff was set up at the Imperial College, London, for lunar computations. The Chapman-Miller method was developed and used there for some years before its publication, at first still taking apparent lunar time as basis. When published in 1940, a referee (Doodson) indicated the desirability of the use of mean lunar time and truly harmonic analysis, and consequently the description was given in terms of Schmidt's regressing lunar phase integers Mu. After World War II Bartels and Chapman decided that it was preferable to use instead of Mu the progressing lunar phase integers Nu or Nu'. At first only the primary harmonic analysis was done by computer; the mechanization of the complete process was first made by Wilkes (1962) in a determination of L in geomagnetic data for San Fernando.

By 1939 the determinations made by handwriting methods had added 28 to the 12 made between 1842 and 1927. Chapman's use of machine methods led to a rapid increase in the number of determinations up to and during World War II, until this brought his computing office to an end; the results thus obtained, for 27 stations, were published in one paper (Chapman and Tschu, 1948). By 1956 Chapman and

Westfold had a total of 69 determinations available for discussion in relation to their geographical distribution and seasonal variations, and also for comparison with $S_2(p)$ results for most of the stations. The determinations of $S_n(p)$ made in the course of the Chapman-Miller method from each subdivision of the often long series of data used to calculate $L_2(p)$, are a valuable by-product of the work.

In the early 1950's Haurwitz, with some of his pupils, took up the study of atmospheric tides and thermal tides, at first more theoretically than observationally. More recently he has become the main contributor, with his colleague Cowley, to our further knowledge of $L_2(p)$ and of other meteorological manifestations of the lunar tide. They have used the Chapman-Miller method, when hourly data have been available. For 7 African stations where only three daily readings were available they used the Bartels method described in Section 2L.8. For ten Australasian stations where trihourly readings were made, except at midnight (thus 7 per day), they used the C-M method, after first calculating A_0 and A_n, B_n for n from 1 to 3, by formulae necessarily different from those usual in harmonic analysis.

The number of determinations of $L_2(p)$ has now risen to 104, as indicated in the Table 2L.2; it gives the annual mean L_2, and also, for 85 stations, the seasonal values for j, e, d. Monthly mean values have been obtained for 13 stations.

The geographical distribution of $L_2(p)$ has been discussed by Chapman and Westfold (1956) on the basis of 69 annual mean values. That study has been much extended by Haurwitz and Cowley (1970), for the seasonal means as well as for the annual mean. Figure 2L.3 (p. 90) gives their map of equilines of the amplitude l_2 for the yearly mean. It has fewer lines than the corresponding map for the annual mean amplitude s_2 of $S_2(p)$, Figure 2S.3, and a smaller part of the earth is covered. It is difficult to compare the geographical distributions of $S_2(p)$ and $L_2(p)$ from such maps. A better method, applied by Haurwitz and Cowley, is to make a spherical harmonic analysis of $L_2(p)$.

2L.10. The Expression of $L_2(p)$ in Spherical Harmonic Functions

Haurwitz and Cowley drew maps, seasonal and annual, showing by equilines the distribution of the harmonic coefficients a_2, b_2 of $L_2(p)$. From these maps the values were read at 24 points round each of 11 circles of latitude 10° apart, from 50°N to 50°S. The sequences of values for each latitude were harmonically analyzed and expressions in terms of waves depending on $2\tau_u + s\phi$ were then derived, as described in Section 2.4A for $S_2(p)$, from the coefficients $\alpha_s(\theta)$, $\beta_s(\theta)$. From the corresponding amplitudes $\gamma_s(\theta)$ a mean amplitude $\bar{\gamma}_s$ was determined for each s, in the same way as for $S_1(p)$; cf. (4) of Section 2S.4B. Figure 2L.4 (p. 90), shows on a logarithmic scale the values of $\bar{\gamma}_s$ thus found. The main wave is that for $s=2$, corresponding to a tide travelling with the moon round the earth. The next greatest amplitudes are about a tenth of the main wave, then come those for $s=0$ and $s=-2$. The waves for which $s \neq 2$ are due to the non-uniform structure of the atmosphere on which the tidal force acts, and on the irregular tidal rise and fall of the surface on which the atmo-

TABLE 2L.2

Amplitude, phase and probable error for the lunar atmospheric tide at 104 stations (Haurwitz and Cowley, 1970). The unit for the amplitudes and probable errors is 1 microbar (μb)

Station	Yrs	Metr.	Colat.	Long.	Ref.	Annual Amp.	Annual Phase	Annual P.E.	d Amp.	d Phase	d P.E.	j Amp.	j Phase	j P.E.	e Amp.	e Phase	e P.E.
1 Oslo	68	25	30.1	10.7	8	6.9	60.8	1.7	12.4	29.0	3.5	5.3	75.1	2.5	7.2	94.4	3.1
2 Uppsala	84	24	30.1	17.6	1	8.5	65.6	1.3	8.6	47.0	3.6	8.8	83.0	2.3	8.8	65.5	2.2
3 Aberdeen	51	14	32.1	357.5	3	15.1	98.0	3.1	*	*	*	*	*	*	*	*	*
4 Glasgow	45	55	34.1	355.8	3	4.8	82.0	3.0	*	*	*	*	*	*	*	*	*
5 Copenhagen	66	13	34.3	12.6	2	10.3	87.5	2.5	7.7	64.3	2.7	12.8	95.3	2.0	11.7	95.0	2.7
6 Keitum	10	11	35.1	8.4	3	8.1	80.0	3.7	*	*	*	*	*	*	*	*	*
7 Hamburg-Potsdam	67	26	36.4	10.0	3	14.3	93.0	2.1	11.9	73.0	3.3	16.8	79.0	3.3	17.6	122.0	4.5
8 Greenwich	64	45	38.5	0.	3	11.7	114.0	2.4	5.7	90.0	4.0	15 5	115.0	3.6	14.4	121.0	4.0
9 Vancouver	12	41	40.7	236.9	3	1.9	58.0	4.4	*	*	*	*	*	*	*	*	*
10 Paris	36	50	41.2	2.5	11	10.1	95.8	3.1	14.3	26.4	5.2	22.4	117.2	6.6	3.1	142.0	7.1
11 Victoria	29	70	41.6	236.7	3	5.3	107.0	2.9	9.0	13.0	4.6	6.0	150.0	2.6	9.0	143.0	5.6
12 Montreal	28	57	44.5	286.4	3	27.0	72.0	4.1	18.0	62.0	7.2	56.0	75.0	5.0	29.0	76.0	8.5
13 Portland	16	47	44.5	237.3	3	2.8	54.0	5.4	14.0	350.5	11.5	7.0	95.8	6.4	8.6	164.3	9.1
14 St. John	17	36	44.7	293.9	3	26.0	83.0	4.7	31.0	88.0	11.9	23.0	95.0	4.7	24.0	66.0	6.5
15 Toronto	31	115	46.3	280.6	3	29.0	78.0	4.1	21.0	71.0	6.6	39.0	89.0	4.1	29.0	70.0	4.5
16 Sapporo	41	17	46.9	141.4	3	25.7	69.0	2.5	26.0	60.0	4.5	28.0	77.0	3.5	24.0	68.0	4.9
17 Tiflis	26	442	48.4	44.5	3	27.0	65.0	2.4	22.0	55.0	4.1	31.0	64.0	4.1	29.0	67.0	4.1
18 Rome	4	50	49.1	12.5	3	27.0	98.0	6.4	*	*	*	*	*	*	*	*	*
19 Naples	10	40	49.2	14.3	12	21.3	74.0	3.1	*	*	*	*	*	*	*	*	*
20 Salt Lake City	16	1329	49.2	248.1	3	7.1	50.5	5.0	17.0	309.4	10.2	13.9	101.8	7.9	17.0	70.8	7.8
21 Coimbra	62	141	49.8	351.6	3	16.9	75.7	1.7	14.0	49.9	3.2	20.3	84.5	2.6	14.0	49.9	3.2
22 Philadelphia	12	22	50.0	284.8	3	28.2	64.2	4.4	22.2	30.1	8.3	37.0	80.5	7.2	29.7	63.9	7.4
23 Santa Cruz	26	40	50.6	328.8	3	18.4	58.0	2.2	22.6	23.0	4.7	19.4	72.0	3.0	19.8	84.0	3.4
24 Washington	12	34	51.1	283.0	3	31.1	58.5	4.9	34.2	12.9	10.3	35.2	75.7	8.2	36.2	76.2	6.5
25 Lisbon	54	95	51.3	350.8	3	22.1	62.4	1.7	23.1	40.0	3.5	22.3	74.5	2.3	23.2	68.3	3.0
26 Lick Obs'y	48	1285	52.1	237.7	3	10.9	67.7	1.6	11 6	19.9	3.0	13.1	87.7	2.6	12.5	87.3	2.8
27 Dodge City	16	766	52.2	260.0	3	21.8	73.3	5.0	28.2	53.4	9.4	32.1	74.9	5.9	8.1	100.9	10.9
28 San Francisco	16	47	52.2	237.6	3	14.7	57.9	2.3	16.3	22.7	3.1	18.9	80.0	3.3	13.5	67.4	5.7
29 Ponta Delgada	39	22	52.3	334.3	3	21.7	55.0	1.6	22.1	52.0	3.3	26.4	63.0	1.9	19.3	70.0	2.9
30 Catania	5	69	52.5	15.0	13	32.8	72.0	3.0	*	*	*	*	*	*	*	*	*
31 San Fernando	56	29	53.5	353.8	3	27.7	79.1	1.5	23.2	50.9	3.1	35.4	90.7	2.3	26.9	83.7	2.6
32 Greensboro	17	397	53.9	280.0	4	36.1	78.3	5.1	39.4	68.5	14.1	39.8	85.7	4.4	30.2	82.3	7.6
33 Tokyo	44	21	54.3	139.8	3	38.6	60.0	2.5	44.0	45 0	4.3	40.0	70.0	3.6	34.0	67.0	5.1
34 Oklahoma City	18	270	54.6	262.4	4	24.3	70.4	8.4	20.5	83.7	14.7	28.5	66.4	11.8	24.3	73.8	10.2
35 Burbank	17	221	55.8	241.6	4	25.8	64.0	3.3	18.4	14.0	7.2	36.7	74.0	4.6	30.0	80.4	4.8
36 Mt. Wilson	7	1782	55.8	241.8	3	15.9	75.5	2.4	23.0	34.9	4.0	10.7	113.7	4.1	21.3	94.2	4.0
37 Kumamoto	38	39	57.2	130.7	3	37.4	48.0	2.2	38.0	29.0	3.7	41.0	52.0	3.8	36.0	62.0	4.1

Table 2L.2 (continued)

Station	Yrs	Metr	Colat.	Long.	Ref.	Annual Amp.	Annual Phase	Annual P.E.	d Amp.	d Phase	d P.E.	j Amp.	j Phase	j P.E.	e Amp.	e Phase	e P.E.
38 San Diego	16	26	57.3	242.8	3	23.2	68.8	3.0	20.3	30.0	5.0	32.8	89.6	4.6	23.4	70.5	5.9
39 Bermuda	6	49	57.6	295.3	3	41.5	66.8	5.9	50.4	17.6	10.8	49.3	87.8	5.7	44.3	88.6	14.1
40 Haifa-Jerusalem	5	10	58.0	35.0	3	23.3	35.6	5.2	23.1	25.1	8.8	29.3	28.6	9.7	16.8	55.1	8.3
41 Helwan	9	30	60.0	31.3	3	36.0	64.0	*	30.0	49.0	*	52.0	65.0	*	27.0	78.0	*
42 New Orleans	12	16	60.0	269.9	3	33.9	69.6	3.9	40.3	44.6	5.8	41.3	92.5	6.2	24.9	67.2	7.9
43 Galveston	12	16	60.7	265.2	3	35.7	70.9	4.9	31.1	32.1	8.2	48.1	85.7	7.9	35.2	79.4	9.2
44 Naha	34	10	63.8	127.6	3	48.6	43.0	2.2	53.0	27.0	3.4	55.0	53.0	4.3	41.0	51.0	3.7
45 Taihoku	36	9	65.0	121.5	3	55.5	50.0	2.1	62.0	22.0	3.2	67.0	63.0	3.6	45.0	67.0	3.8
46 Tamanrasset	8	1380	67.2	5.5	5	29.4	83.5	8.8	29.3	46.5	7.0	36.3	106.6	36.0	32.2	92.3	14.3
47 Hong Kong	28	33	67.7	114.2	3	60.0	60.0	2.4	60.0	35.0	4.1	73.1	69.0	4.1	56.0	73.0	4.1
48 Honolulu	20	15	68.7	202.1	3	45.0	66.0	2.6	48.0	42.0	5.2	48.0	77.0	3.7	44.0	80.0	3.9
49 Mexico City	18	2280	70.6	260.9	3	35.0	74.0	*	36.0	48.0	*	41.0	84.0	*	32.0	89.0	*
50 Wake Island	13	4	70.7	166.6	6	64.0	78.5	18.0	*	*	*	57.0	90.0	*	54.0	92.0	*
51 Bombay	50	11	71.1	72.6	3	49.0	81.0	*	43.0	55.0	4.9	54.7	77.0	4.9	46.4	78.2	4.6
52 Aguadilla	18	67	71.5	292.9	4	48.4	68.7	2.7	47.7	49.1	4.1	61.3	84.3	3.5	52.9	82.2	3.8
53 San Juan	18	20	71.5	293.9	4	50.7	73.2	2.1	43.0	52.9	2.0	87.5	66.1	2.5	69.9	71.5	2.8
54 Manila	45	14	75.4	121.0	3	73.2	59.9	1.4	67.4	34.6	*	62.0	68.0	*	51.0	58.0	*
55 Madras	20	7	76.9	80.3	3	53.0	56.0	*	49.0	39.0	*	73.0	91.8	6.4	68.4	94.6	5.8
56 Willemstad	10	5	77.8	291.0	7	59.8	82.9	2.9	48.3	53.0	4.1	47.0	65.0	*	63.0	68.0	*
57 Periyakulam	7	288	79.1	77.5	3	52.0	62.0	*	36.0	39.0	*	71.6	92.3	5.1	59.8	100.5	5.2
58 Trinidad	16	8	79.4	298.4	7	53.5	90.3	3.2	34.2	69.8	4.3	49.0	80.0	*	61.0	82.0	*
59 Kodaikanal	7	2341	79.8	77.5	3	52.0	68.0	*	38.0	42.0	*	*	*	*	*	*	*
60 Uyelang	4	10	80.3	161.0	8	77.4	95.0	9.1	*	*	*	62.5	71.0	5.9	55.0	86.3	7.3
61 Addis Ababa	7	2443	81.0	321.2	8	60.3	66.4	3.4	73.1	45.8	4.0	72.0	83.8	5.9	60.3	84.7	5.6
62 Balboa	17	6	81.4	280.5	4	58.3	77.0	2.9	45.7	52.9	4.1	*	*	*	*	*	*
63 Agusta	3	1880	81.5	77.3	3	49.3	75.2	7.2	*	*	*	61.0	104.5	9.6	53.4	71.7	10.5
64 Trivandrum	12	60	81.5	77.0	3	50.7	73.1	6.2	63.8	47.2	11.9	78.4	76.7	23.5	76.2	78.8	23.0
65 Lagos	8	7	83.6	3.4	8	69.2	73.1	12.4	52.4	61.9	17.6	*	*	*	*	*	*
66 Yaluit	4	4	84.1	169.6	3	48.4	105.5	8.4	*	*	*	*	*	*	*	*	*
67 Accra	6	100	84.4	359.8	3	45.0	72.0	5.8	*	*	*	*	*	*	*	*	*
68 Singapore	5	5	88.7	103.8	3	89.0	76.0	*	83.5	63.4	25.2	105.6	82.6	31.3	86.0	86.4	26.1
69 Kololo Hill	6	1312	89.7	32.6	3	91.2	77.9	16.0	*	*	*	*	*	*	*	*	*
70 Nauru	6	8	90.4	166.9	8	72.9	98.4	11.1	64.0	90.0	15.0	74.0	109.0	15.0	82.0	100.0	15.0
71 Ocean Is.	6	28	90.9	169.6	5	71.0	100.0	8.0	64.4	64.7	20.1	99.0	76.0	24.5	99.3	75.5	29.6
72 Kabete	6	1829	91.3	36.8	3	87.4	72.5	15.4	*	*	*	*	*	*	*	*	*
73 Neuwied	9	8	92.0	33.0	8	58.1	84.1	8.9	39.4	56.0	5.0	53.5	77.3	7.0	52.2	79.8	5.4
74 Rabaul	11	8	94.2	152.2	9	47.8	73.0	3.8	67.9	47.3	7.7	69.0	86.6	7.1	58.1	85.0	9.0
75 Kwai	5	1610	94.8	38.3	8	61.8	74.6	4.7	42.2	80.8	5.2	62.7	94.9	2.4	52.9	90.3	4.4
76 Tabora	28	1230	95.0	32.9	8	52.2	90.0	2.3	81.1	54.0	3.6	94.0	72.0	3.6	77.1	77.0	3.6
77 Batavia	50	5	96.2	106.8	3	82.0	68.0	2.1	83.9	59.2	26.4	95.8	71.5	28.5	96.0	67.1	28.8
78 Chukhani Palace	6	20	96.2	39.2	3	91.5	65.9	16.1	*	*	*	*	*	*	*	*	*

Table 2L.2 (continued)

Station	Yrs	Metr	Colat.	Long.	Ref.	Annual Amp.	Annual Phase	Annual P.E.	d Amp.	d Phase	d P.E.	j Amp.	j Phase	j P.E.	e Amp.	e Phase	e P.E.
79 Dar es Salaam	9	8	96.8	59.3	8	64.3	83.4	3.1	65.7	81.8	7.4	72.2	88.3	3.3	56.5	77.7	6.6
80 Port Moresby	13	47	99.4	147.2	9	40.2	75.8	3.8	38.8	59.5	5.6	39.2	86.8	5.2	44.8	80.4	4.4
81 Huancayo	15	3355	102.0	284.7	3	39.0	74.1	7.3	27.2	52.0	9.1	52.4	69.0	16.0	41.6	95.0	12.9
82 Lima	7	251	102.0	283.0	3	47.5	73.0	4.6	39.0	49.6	7.5	42.3	81.8	7.5	65.2	79.7	8.7
83 Darwin	13	29	102.5	130.9	9	59.9	49.0	3.1	73.3	35.8	6.2	53.7	49.7	5.3	56.8	65.6	4.8
84 Apia	25	3	103.6	188.2	3	73.0	59.0	*	78.0	56.0	*	80.0	60.0	*	60.0	62.0	*
85 Broken Hill	5	305	104.4	28.4	3	84.4	66.7	15.1	88.8	69.9	27.1	89.0	71.4	26.7	77.0	57.9	24.8
86 St. Helena	5	538	105.9	334.3	3	59.0	93.0	*	*	*	*	*	*	*	*	*	*
87 Broome	12	9	107.9	122.2	9	49.5	72.2	3.6	47.8	65.8	7.1	53.8	71.9	6.1	44.6	77.8	8.4
88 Tananarive	24	1400	108.9	47.5	8	53.4	82.1	1.8	48.9	59.6	4.6	58.0	89.4	2.4	56.3	93.2	3.3
89 Mauritius	40	55	110.1	57.5	3	51.0	98.0	1.1	41.0	76.0	1.7	60.0	106.0	1.7	56.0	104.0	1.7
90 Walvis Bay	12	3	112.9	14.4	8	35.4	99.5	10.4	*	*	*	*	*	*	*	*	*
91 Rockhampton	12	14	113.4	150.5	9	41.7	83.0	3.8	35.5	59.9	6.5	47.9	93.0	5.7	46.5	92.9	5.2
92 Alice Springs	12	548	113.8	133.9	9	38.2	56.4	3.9	39.3	33.2	6.1	46.3	65.4	6.1	32.2	69.2	7.8
93 Pretoria	10	1327	115.5	29.6	3	42.0	89.9	7.9	44.3	79.7	14.9	39.9	90.3	12.2	43.1	100.3	13.5
94 Kimberley	48	1202	118.7	24.0	10	47.3	99.1	1.3	48.8	88.6	2.5	50.6	108.3	2.1	43.8	99.9	2.1
95 Norfolk Is.	12	109	119.0	167.9	9	52.2	71.5	2.9	56.7	55.2	6.4	54.8	79.3	5.3	48.7	83.8	4.8
96 Kalgoorlie	8	361	120.8	121.4	9	41.0	53.8	4.6	47.5	50.3	10.3	40.0	48.2	7.4	36.6	63.9	8.7
97 Lord Howe Is.	4	46	121.5	159.0	9	34.2	71.2	5.1	50.9	52.8	10.3	33.8	99.7	9.0	23.4	73.8	9.4
98 Buenos Aires	20	25	124.6	301.6	3	12.8	87.0	4.2	21.0	44.0	7.0	11.0	113.0	7.0	15.5	121.0	7.0
99 Montevideo	15	29	124.9	304.0	8	19.4	76.1	4.6	26.1	63.9	11.3	17.6	103.8	9.2	17.4	73.2	7.9
100 Carranzo-Tumbes	14	63	126.1	287.1	8	24.1	89.5	8.5	*	*	*	*	*	*	*	*	*
101 Melbourne	24	28	127.6	145.0	3	28.5	84.0	2.3	35.5	79.6	4.7	25.8	100.3	3.1	26.2	73.6	4.0
102 Mocha-Galera	14	28	129.2	286.2	8	29.6	97.9	6.6	*	*	*	*	*	*	*	*	*
103 Wellington	24	3	131.1	174.8	3	38.9	69.5	5.6	37.9	61.0	9.6	39.9	70.2	9.8	39.8	76.6	9.8
104 Hobart	5	55	132.8	147.5	9	24.4	80.1	4.9	38.0	77.6	10.8	12.9	32.5	8.9	29.9	103.1	7.8

* Information unavailable.

REFERENCES TO TABLE 2L.2

(1) Haurwitz, B. and Cowley, A. D.: 1969
(2) Chapman, S.: 1969
(3) Bartels, J., Chapman, S., and Kertz, W.: 1952
(4) Haurwitz, B. and Cowley, Ann D.: 1965
(5) Duclay, F. and Will, R.: 1960
(6) Kiser, W. L., Carpenter, T. H., and Brier, G W.: 1963
(7) Haurwitz, B. and Cowley, Ann D.: 1966

(8) Haurwitz, B. and Cowley, Ann D.: 1967
(9) Haurwitz, B. and Cowley, Ann D.: 1968
(10) Chapman, S. and Hofmeyr, W. L.: 1963
(11) Rougerie, P.: 1957
(12) Palumbo, A.: 1962
(13) Palumbo, A.: 1960

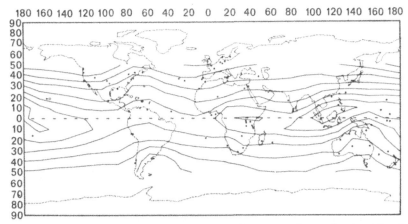

Fig. 2L.3.   World map showing equilines of l_2, the annual mean amplitude of $L_2(p)$. After Haurwitz and Cowley (1970).

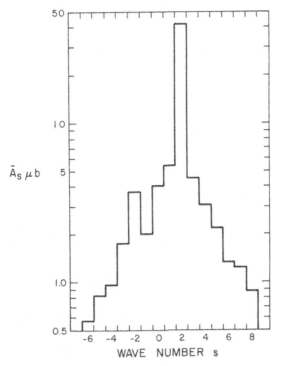

Fig. 2L.4.   The amplitudes (on a logarithmic scale, and averaged over the latitudes from 50°N to 50°S) of the lunar semidiurnal pressure waves, parts of $L_2(p)$, of the type $l_s \sin(2\tau_u + s\phi + \lambda_s)$, where τ_u signifies universal mean lunar time. After Haurwitz and Cowley (1970).

sphere rests; they also include 'noise', partly due to the irregularity and accidental errors of the data.

As described in Section 2S.4B, each wave s (for the five values -2, 0, 1, 2, 3) was expressed in terms of associated Legendre functions of order $|s|$ and degree $n = |s| + i$,

TABLE 2L.3

Main spherical harmonic terms in $L_2(p) = c_{k,s} P_{k,s} \sin(2\tau_u + s\phi + \varepsilon_{k,s})$
Unit 1 μb (Haurwitz and Cowley, 1970)

	Year		Season j		Season e		Season d		$c_k{}^s/c_{k,s}$
	$c_{k,s}$	$\varepsilon_{k,s}$	$c_{k,s}$	$\varepsilon_{k,s}$	$c_{k,s}$	$\varepsilon_{k,s}$	$c_{k,s}$	$\varepsilon_{k,s}$	
$P_{1,1}$	6.9	190°	6.4	155°	5.3	162°	7.4	192	0.866
$P_{2,2}$	57.0	75.2	62.3	78.7	57.5	78.8	52.7	57.3	1.118
$P_{3,2}$	5.7	326.5	3.6	0.9	3.6	265.7	9.1	305.2	1.323
$P_{4,2}$	9.3	225.3	12.7	244.3	10.3	251.6	4.4	193.1	1.5

using the method of weighted least squares; three values of i were found sufficient in each case. Table 2L.3 quotes only those terms in Table 2S.5 that have coefficients (for the normalized functions $P_{n,m}$: cf. Section 2S.4A.1) at least equal to 5 μb for the annual mean $L_2(p)$. The amplitude is denoted by $c_{k,s}$ and the phase by $\varepsilon_{k,s}$; factors are given $c_k^s/c_{k,s}$ by which to multiply $c_{k,s}$ to obtain the amplitude c_k^s of the corresponding Schmidt function P_k^s. Haurwitz and Cowley also give the coefficients $a_{k,s}$ and $b_{k,s}$ corresponding to $c_{k,s}$, and $\varepsilon_{k,s}$, and amplitudes and phases for the Hough functions $s=2$, $k=2, 3, 4$.

For the main term $s=2$, $k=2$ the amplitudes and phases of the Hough functions are very similar to those shown above for $P_{2,2}^2$.

To test how well the various analytical expressions derived for $L_2(p)$ agree with the original equiline maps for a_2, b_2, Haurwitz and Cowley calculated an overall index of the differences between the originally measured values of a_2, b_2 (themselves a smoothed version of the irregularly distributed data for $L_2(p)$), and the values calculated from the formulae; the method used resembles that indicated by (5) of Section 2S.4B. Comparisons of 3 kinds were made, using 'analytical' values calculated (i) from the three Hough functions above mentioned; (ii) using only the three Legendre functions for $s=2$, $k=2, 3, 4$; and (iii) using all 15 terms of their table. This was done for the year and the three seasons, y, j, e, d. For (i) and (ii) the results were very similar, approximately 11 μb in all 8 cases; for (iii), depending on many more adjusted constants, the 'error' was reduced by 20 to 30%. It was also found that only insignificant changes resulted if the 'errors' were calculated using weighting factors, as in (4) of Section 2S.4B, proportional to the area of the different latitude zones.

In all four cases, y, j, e, d, the 'error' is about $\frac{1}{5}$ of the amplitude of the principal term $P_{2,2}$. For $S_1(p)$ the corresponding ratio was larger, $\frac{1}{3}$ of the amplitude of the principal term $P_{1,1}$. This means that the irregular features superposed on the principal travelling wave are proportionately greater for $S_1(p)$ than for $L_2(p)$.

2L.11. The Asymmetry of $L_2(p)$ Relative to the Equator, and its Seasonal Variation

Figures 2L.5a, b, c, d, respectively (Haurwitz and Cowley, 1970) are harmonic dials

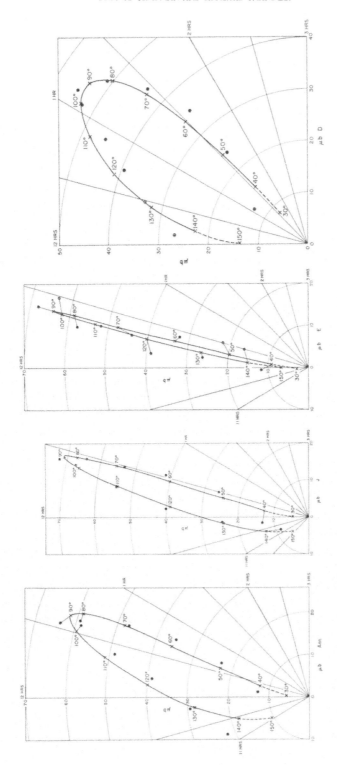

Fig. 2L.5. Harmonic dials showing $L_2(p)$ for the mean of the year (Figure 2L.5a) and for the seasons j, e, d (Figures 2L.5b, c, d respectively), as a function of colatitude from 30° (60° N) to 150° (60° S). The dots give the values obtained by harmonic analysis of the a_2, b_2 read from maps of equilines. The crosses on the continuous curves indicate the values computed from three spherical harmonic terms, $P_{k,s}$, $s = 2$, $k = 2$, 3, 4. Corresponding dots and crosses are joined where their identity might seem uncertain. After Haurwitz and Cowley (1970).

for $L_2(p)$ for the year (y) and the three seasons j, e, d. The dots show the components α_s, β_s for each colatitude θ from 30° to 150°, derived as explained in Section 2L.10 (or Section 2S.4A); the crosses show the values computed from the sum of the three terms $s=2$, $k=2$, 3, 4 of Table 2S.5. The crosses are joined by continuous curves. The dials show that except for the season e there is considerable asymmetry between the northern and southern hemispheres. This was noted earlier, from data for fewer stations, especially in the south, by Chapman and Westfold (1956), as shown by the curve l_2 of Figure 2L.6. The Figures 2L.5a, b, d for y, j, d show that at southern stations the maxima of $L_2(p)$ occur earlier than at the same latitudes in the north. The amplitudes are greater, for the same distance from the equator, in the south than in the north.

Figures 2L.5b, c, d also well show the very considerable seasonal change of $L_2(p)$, both in amplitude and phase, in the different latitudes. This is shown in more detail (monthly) in Figure 2L.7, consisting of harmonic dials for $L_2(p)$ for 8 stations individually or averaged (nos. 45, 21+25+31, 7, 77, 47 of Table 2L.2). All the stations are northern except Batavia, which shows a similar variation throughout the year, as at the northern stations. The change of phase from January to June is still greater than is shown for the 4-monthly seasons by Figure 2L.5.

Haurwitz and Cowley (1969) discuss the possible causes of the asymmetry and the seasonal variation. The great difference between the ocean area in the northern and southern hemispheres will certainly affect whatever part of $L_2(p)$ is due to the lunar tidal rise and fall of the surface underlying the atmosphere. An investigation by Sawada (1965) of the influence of the oceanic rise and fall upon $L_2(p)$ suggests that the hemispheric asymmetry may produce an asymmetry of $L_2(p)$ in the observed direction. There will also be considerable asymmetry in the surface friction over the two hemispheres, for the same reason; namely, the land-sea disparity between them. This will also affect the temperature-height distribution over them, and its seasonal variation.

2L.12. Comparison of $L_2(p)$ and $S_2(p)$

Figure 2L.6 indicates the variation with latitude, from north to south, of the annual mean amplitudes l_2 and s_2 of $L_2(p)$ and $S_2(p)$, averaged over 10° belts of latitude, derived from 69 stations. There is a 20-fold ratio of the scales for L_2 and S_2. The considerable similarity and overlap of the S_2 and L_2 graphs shows that the ratio s_2/l_2 is close to 20.

This ratio can be inferred more exactly, on the basis of considerably more data, both for S_2 and L_2, from the expressions for the main term P_2^2 in their spherical harmonic expressions. For S_2 Haurwitz (Section 2.4A) gives

$$1.16 \text{ mb} \sin^3 \theta \sin(2t + 158°).$$

For L_2 the main term found by Haurwitz and Cowley is $57.0 P_{2,2} \sin(2\tau + 75°)$, or,

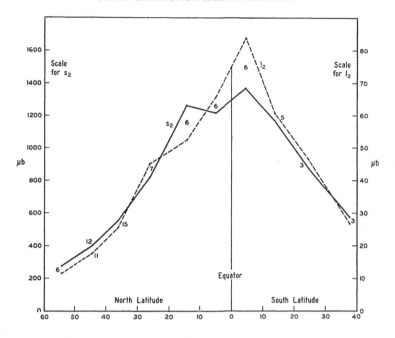

Fig. 2L.6.   Mean values of the amplitudes s_2 (full line) and l_2 (broken line) of the annual mean solar and lunar semidiurnal air-tides in barometric pressure, $S_2(p)$ and $L_2(p)$, for $10°$ belts of latitude. The numbers beside each point show from how many stations that point was determined. After Chapman and Westfold (1956).

because $P_{2,2} = 0.9684 \sin^3 \theta$, it is

$$55.2 \ \mu\text{b} \sin^3 \theta \sin (2\tau + 75°).$$

Thus $s_2/l_2 = 1160/55.2 = 21.0$. The phase difference $\sigma_2 - \lambda_2 = 83°$.

These results are further indicated in more detail by Figures 2L.8a, b, which show the dial vectors of the annual mean $S_2(p)$ and $L_2(p)$ for many places widely spread over the globe. Again the scales for S_2 and L_2 differ by a factor of 20. The probable error circles for L_2 are shown. Clearly on the whole the vectors for S_2 and L_2 on their different scales are about equal, and their phases differ by slightly less than $90°$, though both as regards amplitude and phase there are striking exceptions.

A remarkable feature indicated by comparison of Figures 2S.1 and 2L.7 is the great difference between the seasonal variations of $S_2(p)$ and $L_2(p)$. It is proportionately much greater for $L_2(p)$ than for $S_2(p)$, exactly the opposite to what one would expect. The thermal excitation of $S_2(p)$ must certainly differ considerably from summer to winter, whereas the tidal L forces have no seasonal variation. The considerable seasonal variation of $L_2(p)$ must indicate some notable disparity of atmospheric structure between summer and winter. It would seem that this change partly cancels the influence of the change of excitation of S_2.

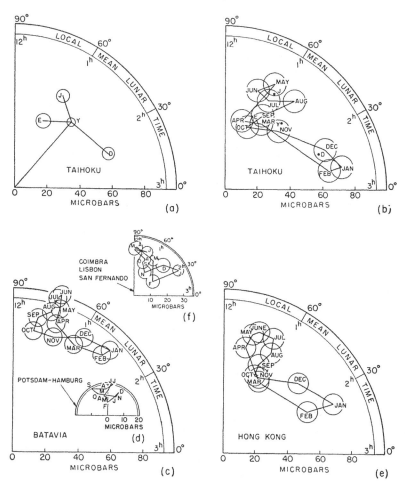

Fig. 2L.7. Harmonic dials, with probable error circles, indicating the changes of the lunar semi-diurnal air-tide in barometric pressure in the course of a year, (a) Annual (y) and four-monthly seasonal (j, e, d) determinations for Taihoku, Formosa (now Taipei, Taiwan) (1897–1932). Also five sets of twelve monthly-mean dial points. See Table 2L.2 for particulars of the seven stations. After Chapman (1951).

2L.13. The Lunar Tidal Wind Variation

The tidal variation of sea level is associated with tidal currents, superposed on any other motions present, such as the Gulf Stream. Similarly in the atmosphere the $L_2(p)$ variations must involve lunar tidal *wind* variations. They will be small, and are overlaid by much larger winds, which are very variable, changing direction sometimes many times within a single hour. Hence it is to be expected that it will be far more difficult to determine the lunar tidal wind variation $L(\mathbf{V})$ from wind data than it is to determine $L(p)$ from barometric data. Wind data are usually registered and published in the form of wind speed and wind direction. To determine

$L(\mathbf{V})$ it is necessary to convert the data into hourly values of the wind components, u southward and v eastward. Only very few observatories have given their wind data already converted in this way; among them are Mauritius and Bombay. The first attempt to determine $L_2(u)$ and $L_2(v)$ was based on the Mauritius data (Chapman, 1948) for about 16 years, 1916, 1917 and 1920–33. The results obtained were not well determined. Recently Haurwitz and Cowley (1968, 1969) have made six more determinations, for the stations 32, 47, 52, 53 and 62 of Table 2L.2 and Uppsala (59.9 °N, 17.6 °E). Their original data were in the usual form of wind speed and direction. With modern computers the labor of conversion to u and v, once the data are on cards or tape, is not serious. Some of their data covered periods of about 20 years, but for Uppsala and Hongkong 84 and 67 years respectively were used. Again the results were not well determined, apart from the notable exception of Hongkong. All the results are shown in Table 2L.4.

TABLE 2L.4

Lunar tidal wind components (unit 1 cm/sec)

	Period	Latitude	Days	Southward component u			Eastward component v		
				l_2	P.E.	λ_2	l_2	P.E.	λ_2
Uppsala, Sweden	1874–1957	59.9°	N 25752	0.64	0.37	204°	0.75	0.27	179°
Greensboro, N.C.	1945–62	36.1	N 6182	1.3	1.0	191	1.8	1.0	80
Hongkong	a	22.3	N 22892	1.0	0.4	98	2.2	0.4	69
San Juan, P.R.	1941–62	18.5	N 7650	1.4	0.7	267	0.6	1.0	253
Aguadilla	1945–62	15.5	N 6203	1.5	0.9	245	1.5	0.9	100
Balboa, Panama	1941–61	9.0	N 7435	1.2	0.6	209	0.6	0.8	195
Mauritius	1916, 17, 20–23	20.1	S 5730	1.2	0.6	176	1.0	0.6	220

[a] 1890–1939, 1947–59, 1963–66, omitting days when the daily range of either component exceeded 10 m/sec.

Six of these fourteen determinations have amplitudes that are at least *twice* the probable error; this corresponds to nearly 95% significance. But only for $L(v)$ at Hongkong is the ratio at least 3 (actually 5.5), the value hitherto generally regarded as a satisfactory minimum. In two of the eight cases where the ratio is less than 2, it is actually less than 1. Clearly there is need for more reliable determinations, requiring longer series of data. In all the above cases, the C-M method was used. Haurwitz and Cowley omitted some days of unusual or too irregular wind strength; if too many are omitted, however, the reduction in the number of days increases the

Fig. 2L.8.   Harmonic dial vectors for the lunar and solar semidiurnal air-tides $L_2(p)$ and $S_2(p)$ in barometric pressure at 69 stations (omitting S_2 for station 3 and L_2 for station 8). The lunar amplitudes l_2 are on a scale 20 times that for the solar amplitudes s_2. The circles or dots surrounding the 'free' ends of the L_2 vectors indicate the probable errors of L_2. The vectors refer to the points at their common origin. After Chapman and Westfold (1956); the numbers alongside each point refer to their list of stations.

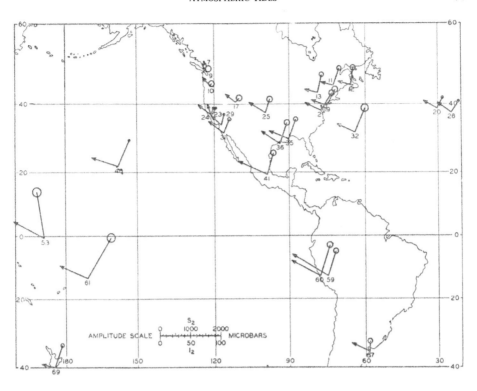

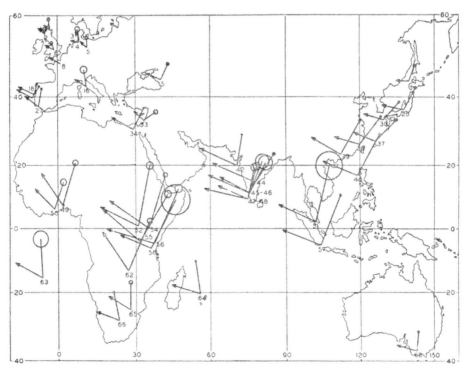

probable error by more than the reduction coming from the smaller error per day. It seems that at least 60 years data are needed for places where the wind regime is like that at the seven stations of the Table.

Chapman illustrated his results for Mauritius by Figures 2L.9 and 2L.10. The former gives harmonic dials showing $L_2(-u)$ and $L_2(v)$ and also, on a tenfold

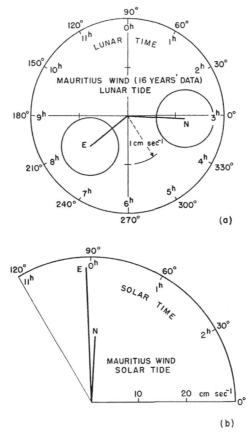

Fig. 2L.9.   Harmonic dials showing (above) the L_2 vectors (with probable error circles) for the northward and eastward lunar tidal wind components at Mauritius, based on about 16 years' hourly data; and (below) dials showing the corresponding S_2 dial vectors. The vectors of the upper diagram are magnified ten times compared with those in the lower diagram. After Chapman (1949, 1951).

different scale, $S_2(-u)$ and $S_2(v)$, determined by the C-M method as a by-product in finding L_2. For L_2 the probable errors are shown; they are notably larger, proportionately, than in the other harmonic dials here given (Figures 1.3, 2L.7, 2L.8). The amplitudes of $L_2(-u)$ and $L_2(v)$ are similar, their phases are considerably different; for S_2 the phases are nearly the same, but $S_2(v)$ has nearly twice the amplitude of $S_2(-u)$.

Figure 2L.10 shows the magnitude and direction of the lunar tidal wind V_L at each lunar hour, and likewise (on a tenfold different scale) of the solar thermal tidal wind V_S at each solar hour. These vector end points describe an ellipse twice daily. These ellipses can also bear another interpretation: they show the path of an air particle from its mean position, corresponding to V_L and V_S; for this interpretation the length scale is shown to the left of the origin, and the hour numbers must be

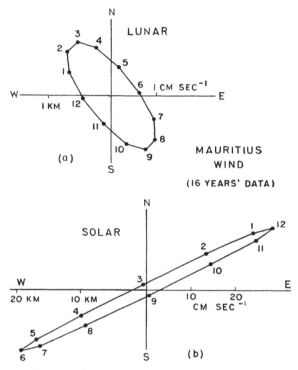

Fig. 2L.10. Vectograms based on Figure 2L.9: they show in plan the wind velocities at each mean lunar or solar hour, morning and afternoon, associated with the lunar and solar semidiurnal wind variations at Mauritius. The speed at each hour is represented (on the scale shown on the right of the origin) by the distance from the dial point (at that hour) to the origin. The diagrams also illustrate the corresponding paths of an air particle at Mauritius, due to these wind variations, if the hour numbers are increased by three. The distance scales are shown to the left of the origin. There is a scale ratio of 10 between the two diagrams. After Chapman (1949, 1951).

increased by 3. The ratio of the speed scale, shown to the right of the origin, to the length scale, is $T/4\pi$, where T denotes respectively the lunar or solar day in seconds. The extreme departures of the air particle from its mean position at Mauritius, in the absence of other winds, is about 23 km for S and about 1 km for L.

In 2S.6 a comparison was made between the observed values of $S_2(V)$ at Uppsala, Hongkong and Mauritius, and what theory would infer from $S_2(p)$, taking account of the earth's rotation, but neglecting friction. A similar comparison relating to $L_2(p)$ and $L_2(V)$ is here made. In the lunar case,

$$\frac{\partial}{\partial t} = \frac{2\pi}{T_L} \frac{\partial}{\partial \phi} = \frac{T_S}{T_L} \omega \frac{\partial}{\partial \phi}.$$

The ratio T_L/T_S is 1.0335; if this factor, for a first approximation, is treated as unity, then the solutions given in 2S.6 for $S_2(u)$ and $S_2(v)$ apply also to $L_2(u)$ and $L_2(v)$, if there we substitute for p_S and σ the values p_L and λ, given by Haurwitz and Cowley (1968) as 55.2 μb and 75° (see 2L.12), and also C_L ($=0.95$ cm/sec) for C_S. Thus for Uppsala, Hongkong and Mauritius respectively the theoretical amplitudes and phases are 1.9, 0.9, 0.8 and 165°, 165°, 345° for $L_2(u)$, and 1.9, 1.2, 1.2 and 235°, 235°, 235° for $L_2(v)$. The amplitudes are of the same order of magnitude as the values derived from observation, but the latter do not have the systematic increase with latitude; the Uppsala determined value is only one third of the theoretical value, whereas for the best-determined result, that for $L_2(v)$ at Hongkong, the determined value is almost twice the theoretical value. There is also much disagreement between the two sets of phases.

In Figure 2L.10 the direction of rotation of the wind vector of $L_2(\mathbf{V})$ is clockwise; this corresponds (see 2S.6) to the negative value of $\lambda_u - \lambda_u$ for Mauritius; for all the northern stations $\lambda_v - \lambda_v$ is positive, so that the rotation in similar diagrams for their $L_2(\mathbf{V})$ would be counterclockwise. All these directions are opposite to what the equations of motion lead to. This adds another to the problems yet to be solved in bringing theory into accord with the observational data for the lunar atmospheric tide.

There is a possibility that $L_2(\mathbf{V})$ might be found more readily, with less or no more data than was used in the above determinations, by using data from stations where (at least at certain seasons) the wind is specially gentle and regular. This is the case, for instance, at Fairbanks, Alaska, during most winter days, when usually there is a great inversion of temperature. Ordinary anemometers fail to measure the wind when the speed is less than 2 knots; according to an unpublished study by Fogle, there were 1176 hours per year, over the average of the period 1959–61, when for this reason there was no wind record there. More sensitive anemometers are now available that could measure the wind during such calm periods. From such data $S(\mathbf{V})$ should be determinable for that season after a year or two, and $L(\mathbf{V})$ from twenty years' data, or perhaps from ten years.

2L.14. The Lunar Tidal Variation of Air Temperature

The heating of the atmosphere by moonlight is quite negligible; if it were otherwise there would be a lunar *diurnal* variation of air temperature, $L_1(p)$. But the moon does produce a lunar *semidiurnal* air temperature variation, as a secondary consequence of its mechanical tidal action. The changes of air density accompanying the tidal variation of pressure will depend on whether the pressure changes adiabatically or isothermally, or something between these extremes. A similar question arises in connection with sound waves. Newton, who was the first to try to calculate the speed of sound, assumed that the density changes are isothermal, and his result did not agree with his measurements. Laplace realized that the density variations are too rapid for the heat of compression to be conducted away during the brief period of each oscillation; the assumption that the changes are adiabatic led him to the correct formula for the speed of sound.

The lunar atmospheric tide is a double tidal wave travelling round the earth each lunar day. The period is long – half a lunar day; but the distance between the regions of compression and rarefaction, or high and low pressure, is also great, except in high latitudes. Calculation (Chapman, 1932b) shows that the horizontal conduction is quite negligible. Consider for simplicity a uniform medium in which there is generation of heat of amount

$$a \sin(pt + 2\pi x/\lambda),$$

where x is the horizontal coordinate. The equation of horizontal conduction is

$$\varrho\sigma \frac{\partial T}{\partial t} = k \frac{\partial^2 T}{\partial x^2} + a \sin\left(pt + \frac{2\pi x}{\lambda}\right).$$

Here σ denotes the specific heat and k the thermal conductivity. If the first term on the right, corresponding to the horizontal conduction, is omitted,

$$T = T_0 + \beta \cos(pt + 2\pi x/\lambda), \qquad \beta = - a/\varrho\sigma p.$$

The term $k \, \partial^2 T/\partial x^2$ is then of order $4\pi^2 k\beta/\lambda^2$, and its ratio to the other two terms is $4\pi^2 k/\varrho\sigma p\lambda^2$. For lunar semidiurnal variations $p = 2\pi/43350 = 1.45 \times 10^{-4}$/sec; for air $\varrho = 1.25 \times 10^{-3}$, $\sigma = 0.24$, and k (the eddy conductivity, much greater than the molecular conductivity except at great heights) may be taken as 15 to 30, according to the wind speed and roughness of the ground: over the sea it appears to be smaller (Taylor, 1917). Taking the larger value of k, the ratio is $2.72 \times 10^{10}/\lambda^2$. At the equator the wavelength λ is half the circumference of the earth, or 10^9 cm; in latitude $L°$ it is $10^9 \cos L°$. Thus the ratio is $2.7 \times 10^{-8}/\cos^2 L°$. This is quite negligible up to very high latitudes (e.g., at latitude 80° it is 3×10^{-5}).

Ignoring horizontal conduction, we next consider vertical conduction. The factor k/ϱ in the above ratio increases rapidly with height, but neglecting this increase for low levels near the ground, the most important change in the ratio comes from the great reduction in the value of λ. For the lunar atmospheric tide λ is greater than 10 km or 10^6 cm, and hence the above ratio is less than 1%. At higher levels, however, thermal conduction will become important. Chapman (1932b) suggested 120 km as such a height, but this may be an overestimate.

It remains to consider conduction from the atmosphere to the medium below, land or sea. Chapman (1932b) considered conduction from an upper to a lower uniform media; in the upper one (ϱ, σ, k) there is a periodic generation of heat $a \sin pt$ per unit volume, which causes heat to flow into the other medium (ϱ', σ', k'). If there were no such flow, T in the upper medium would be as stated above; the presence of the lower medium reduces the interface temperature amplitude β in the ratio

$$k\mu/(k\mu + k'\mu'), \qquad \mu^2 = p\varrho\sigma/2k, \qquad \mu'^2 = p\varrho'\sigma'/2k'.$$

In the atmosphere near the ground, taking $k = 30$ over land, and 1 over the sea, the corresponding values of $k\mu$ are 8.1×10^{-4} and 1.5×10^{-4}. For the land we may take $\varrho\sigma$ to be about 0.4 (e.g., $\varrho=2$, $\sigma=0.2$); k depends on the nature of the ground, e.g., for dry sandstone k is about 10^{-4}, and 6×10^{-3} for granite: for moist ground it may be greater. The reduction of the interface temperature amplitude in the first case is by a factor about 0.95, and in the second, about $\frac{2}{3}$.

Over the sea, where in the air $k\mu$ is 1.5×10^{-4}, the greater conductivity of the sea water may reduce the interface temperature variation more drastically. As in the air, the molecular conductivity is much smaller than the eddy conductivity, which varies with the wind strength; for moderate winds (10 m/sec), $k\mu$ appears to be about 300. For water $\varrho'\sigma'=1$ ($\varrho'=1$, $\sigma=1$). Taking $k=1$, the reduction of β is found to be by a factor 3×10^{-3} – that is, in the case of $L_2(p)$, the adiabatic variation is almost completely annulled; the pressure variations are isothermal.

The adiabatic nature of the tidal air wave propagation has been tested (Chapman, 1932a) by determining the lunar semidiurnal variation of air temperature at Batavia, from 62 years' hourly observations. Figure 2L.11 shows the resulting dial vector

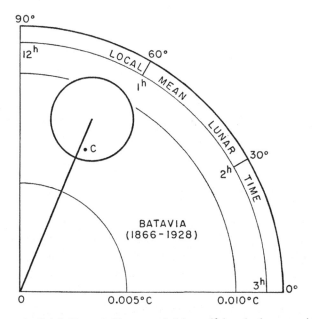

Fig. 2L.11. Harmonic dial (with probable error circle) specifying the lunar semidiurnal variation of air temperature at Batavia. The point C represents the variation calculated from the lunar semi-diurnal barometric variation at Batavia, on the assumption that the variation of pressure and density are adiabatic. After Chapman (1932a, 1951).

with its probable error circle. The point C shows the dial vector calculated from $L_2(p)$ at Batavia on the assumption that the tidal pressure changes are adiabatic; the agreement is within the probable error.

Chapman discussed the situation of Batavia relative to the sea, and the wind regime there, and concluded that it is effectively a land station, where the reduction of an adiabatic pressure change of temperature would at most be of order 10%.

It would be of interest to determine the lunar semidiurnal variation of air temperature from a long series of hourly observations made on the windward side of some small flat tropical island in a great ocean. This would throw light on the degree of interchange of heat between the air and the sea. Bartels suggested that it might prove advantageous in such a study to use only the night record, if this was less variable than the day record.

In the study of $L_2(T)$ at Batavia a subsidiary reduction was made for the rainless days, about 50% of the whole (but many more in the j than in the other seasons); but this did not lead to any improvement of the result shown in Figure 2L.11.

2L.15. The Lunar Tidal Changes of Height of Various Pressure Levels

The lunar atmospheric tide heaps up the air over certain regions and reduces its amount over other regions, and these regions change as the moon revolves relative to the rotating earth. The tidal changes of pressure at ground level are accompanied by changes at higher fixed levels; this means that the isobaric or constant-pressure surfaces rise and fall during each lunar half-day. The first direct evidence of this was provided by Appleton and Weekes (1939). They found a lunar semidiurnal variation of the noon height of the E-layer above southern England, amounting to a twice (lunar) daily rise and fall of about 1 km from the mean level. Their result is illustrated by Figure 2L.12, which shows 11 dial points, each representing a determination from a half month (12 to 14 days), between August 1937 and July 1938. The cross shows the mean dial point, and the probable error circle shows the uncertainty of any one of the 11 dial points. The radius of the probable error circle for the mean dial point is less by a factor slightly greater than 3, so that the determination is a good one.

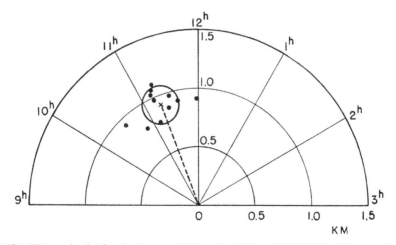

Fig. 2L.12.   Harmonic dial for the lunar semidiurnal variation of the height of the E-region of the ionosphere near London, England. The circle shows the probable error for any one of the eleven separate dial points, each based on 12–14 days' data. After Appleton and Weekes (1948).

Since 1939 there have been many studies of lunar tidal influences on the ionosphere, of two main kinds, namely of the layer heights, for the D-, F_1- and F_2-layers as well as for the E-layer: and of the electron density of each of these layers. This electron density is influenced by the lunar tide in a complicated way; it is determined by a changing balance between processes of electron detachment by absorption of solar photons of short wavelength, and of recombination or charge exchange. The lunar tidal changes of height and electron density affect other ionospheric properties also, such as the absorption of solar photons of particular wavelengths. Reports on many of these studies of lunar tidal influences on the ionosphere up to 1963 are given in the bibliography mentioned at the beginning of the References.

The different ionospheric layers, or their levels of peak density, are not isobaric surfaces, so that the deduction from them of the pressure changes at particular heights is a somewhat involved matter.

One notable feature of the results is that the amplitude of rise and fall of the successive ionospheric layers, D, E, F_1, F_2 increases with height, as the tidal theory of Chapter 3 would imply.

Evidence of pressure changes at a lower level between the ionosphere and the ground, or of height changes of an isobaric surface there, was given by Duperier (1946). Cosmic ray observations provide, in a surprising and most interesting way, information as to the lunar air tide at a level of 18 or 20 km above the ground. Mesons are generated by primary cosmic rays at these levels, and those that descend to the ground can be recorded there. As they are unstable, a proportion of them are transformed on their way to the ground. If the lunar tide raises or lowers the level of the mean air pressure at which the mesons are generated, their path to the ground will be lengthened or shortened. This will cause more or fewer to be transformed in the air, and will reduce or increase the number of survivors recorded at ground level. Duperier (1946) made a determination of L_2, with satisfactory probable error, in his recorded amounts of cosmic ray reception (mainly mesons) at London. From his result he inferred that the lunar tide at about 18 km is considerably magnified (about tenfold what the equilibrium tide would indicate). Studies of this kind deserve more attention and effort, both observational and theoretical.

2L.16. Brief Mention of the Lunar Geomagnetic Tide

Nearly a century ago a small lunar daily variation was detected in the records of the components (or 'elements') of the earth's magnetic field. This variation is produced (like the corresponding solar daily geomagnetic variation) mainly above the earth's surface, though these varying 'primary' magnetic fields of external origin induce electric currents in the conducting body of the earth – mainly deep down, but also, to a lesser degree, near the surface, where they can be measured and recorded. The analysis of these earth-current records reveals solar and lunar daily variations, which form yet another curious by-product, as in the cosmic rays, of the high-level solar and lunar tidal atmospheric oscillations.

These layers are not yet clearly identified, but the E-layer is probably the main region of flow. The flow is probably chiefly in the E-layer, but from time to time solar flares enhance the electron density of the D-layer, and the electric currents flow there also until the flare ceases and the extra D-ionization dies away. There is some electric current flow; however, attaining a far higher level above the equator, because it goes from the northern to the southern hemisphere, or vice versa, along the geomagnetic field lines.

The currents are generated, as first suggested by Balfour Stewart, as in a dynamo. The moving air corresponds to the armature, the conducting ionospheric layers to the armature windings, and the earth's magnetic field to that of the dynamo pole pieces.

The air motion effective for dynamo action is mainly horizontal. The part due to the solar action, mainly thermal, varies in the course of a solar day, and the part due to the moon's tidal action varies in the course of a lunar day. This results in corresponding daily geomagnetic variations, S and L. The L-variation is lunar semidiurnal when averaged over a lunation, but at any particular lunar phase the lunar-induced electric currents are stronger over the sunlit than over the dark hemisphere, because of the greater ionization over the former hemisphere. Consequently the magnetic records show not only a lunar semidiurnal component L_2, but also lunisolar components L_n, for which $n = p - 2/M$, p being an integer (Chapman and Bartels, 1940, Chap. 8).

From the determinations of S_n and L_n in the magnetic records of many stations it is possible to determine the distribution and intensity of these solar and lunar daily-varying electric current systems in the ionosphere, and also the *type* of the inducing atmospheric motions at those levels. To infer the *intensity* of these motions requires a knowledge of the electric conductivity of the layers in which the known electric currents flow. Until the precise situation of the currents is ascertained, and their electric conductivity, the intensity of the solar daily and lunar daily oscillations in the ionosphere cannot be precisely inferred from the geomagnetic data. The present indication is that the lunar tidal horizontal movements, like the lunar tidal rise and fall of the E-layer, are very greatly magnified as compared with what the barometric L_2 data would suggest. It is, however, of interest to note that the ratio of S_2 to L_2 in the magnetic records is about the same as that in the barometric variations.

As mentioned in Section 2S.9C, calculations of the wind distributions responsible for the geomagnetic S and L have been made (Maeda, 1955; Kato, 1956); naturally they simplified the problem considerably in order to arrive at definite results. The complexity of the problem has been stressed by Price (1969). One apparently clear result has been obtained, namely, that the lunar air flow in the layer where the geomagnetic L is generated is opposite to that near ground level. There is much scope for further study, both by observation and theory, of the bearing of these geomagnetic data on the solar and lunar atmospheric tides and thermal tides.

CHAPTER 3

QUANTITATIVE THEORY OF ATMOSPHERIC TIDES
AND THERMAL TIDES

3.1. Introduction

The theory of atmospheric tides and thermal tides has two main parts: (a) The investigation of the sources of periodic excitation, and (b) The calculation of the atmospheric response to the excitation. The former could include a detailed consideration of atmospheric composition and chemistry, the solar spectrum, molecular absorption, radiative transfer, turbulent transfer, and other topics. A study of (b) could in principle concern all the problems involved in the general circulation of the entire atmosphere, including non-adiabatic, orographic, non-linear, hydromagnetic, and numerous other types of process. It is not possible to be so comprehensive in this review, and we here concern ourselves with such topics, as have previous reviewers (Wilkes, 1949; Chapman, 1951; Kertz, 1957; Siebert, 1961; Craig, 1965), only to the extent that they have been used in current investigations. We begin by describing the equations for the atmospheric response to arbitrary excitations – invoking numerous assumptions and approximations. Some attention is given to methods of solution for the equations obtained. We then describe, rather simply, the sources of excitation included in various tidal calculations, and the specific responses inferred. Where possible, the effect of various approximations in these solutions is discussed.

3.2. Equations

In this section we deal with the response of the atmospheric pressure, density, temperature and velocity, but not with the geomagnetic response. Among the approximations used, the following are almost universally applied without much question:

(a) The motion of the atmosphere may be described by the Navier-Stokes equations for a compressible gas. It is convenient to express them in spherical coordinates for a frame of reference rotating with the earth. It is rare to find these equations written down in full. However, some idea of their appearance may be obtained from Goldstein (1938), Haurwitz (1951), and Brunt (1939).

(b) The atmosphere is always in local thermodynamic equilibrium; i.e., it responds to heating via a continuous sequence of equilibrium states. This matter is discussed further by Landau and Lifshitz (1959) and Roberts (1967). Somewhat less rigorous assumptions are:

(c) The atmosphere is a perfect gas, so that, with the usual symbols (see p. 3),*

$$p = \varrho RT. \tag{1}$$

* The most common symbols and their meaning as used in this chapter are listed on pp. 172–174.

Moreover, it is assumed that the atmosphere is of constant composition, so that R is a constant. This is nearly true up to about 95 km (Nawrocki and Papa, 1961). $R = 2.871 \times 10^6 \, \text{erg} \, g^{-1} \, \text{deg}^{-1}$ at the ground, $2.878 \times 10^6 \, \text{erg} \, g^{-1} \, \text{deg}^{-1}$ at 100 km and $3.16 \times 10^6 \, \text{erg} \, g^{-1} \, \text{deg}^{-1}$ at 200 km.

(d) The atmosphere is regarded as a geometrically thin fluid layer of small thickness compared with a, the radius of the earth. Thus, expressing the distance from the earth's center as

$$r = a + z, \tag{2}$$

we neglect terms in our equations whose order is z/a. One consequence is that the acceleration of gravity, g, is treated as a constant; at $z = 100$ km the error is only 3%. If the shallow atmosphere approximation is introduced into the definition of the metric factors for spherical polar coordinates, then the neglect of the component of the earth's rotation vector parallel to the earth's surface follows (Phillips, 1966).* The physical validity of this procedure appears to depend on a wave's frequency being much less than the Brunt-Väisälä frequency (Phillips, 1968).

(e) The atmosphere is taken to be in hydrostatic equilibrium. This implies that vertical accelerations are small compared with g, so that

$$\frac{1}{\varrho} \frac{\partial p}{\partial z} = -g. \tag{3}$$

Solberg (1936) felt concern over this approximation, but Hylleraas (1939) found the approximation to be good, and *a posteriori* checks of obtained solutions appear to confirm this (Yanowitch, 1966).

(f) The earth's ellipticity is ignored.

None of the above seems to constitute serious limitations on atmospheric tidal theory. The following are more significant approximations:

(g) The earth's surface topography is ignored, so that the influence of mountains and the land-sea distribution is not taken into account.

(h) Dissipative processes such as molecular and turbulent viscosity and conductivity, ion drag, and infrared radiative transfer are ignored.

(i) Tidal fields are considered as linearizable perturbations about some basic state. Let a given field be written

$$f = f_0 + f',$$

where f_0 is the basic field and f' is the tidal contribution to f. By 'linearizable' we mean that we may neglect quadratic and higher order terms in f'. The detailed procedure of linearizing the equations of hydrodynamics is given by Haurwitz (1951) and is reproduced by Craig (1965). If there is a tidal excitation, E, then f' is proportional

---

* The equations of motion have been expressed by Hough (1897) and Phillips (1966) in terms of general curvilinear coordinates. For spherical polar coordinates the metric factors are $h_\phi = r \sin\theta$, $h_\theta = r$ and $h_r = 1$. The 'shallow atmosphere' approximation consists in replacing these by $h_\phi = a \sin\theta$, $h_\theta = a$ and $h_r = 1$ (r = distance from center of the earth, θ = colatitude, a = radius of the solid earth).

to E, and linearization requires at the least that E be sufficiently small. As we shall see, however (Section 3.6.C), this is not enough.

(j) Even linear equations can become intractable if their coefficients are at all complicated. In order to make the tidal equations tractable, a number of approximations are made for the basic fields. First, they are assumed to be steady; this should be adequate if the non-tidal fields change with time scales much longer than tidal periods. Second, we assume that the basic flow may be set equal to zero. This, in turn, implies that T_0, p_0 and ϱ_0 are independent of latitude and longitude. From Equations (1) and (3) we then obtain

$$p_0 = p_0(0)e^{-x} \qquad (4)$$
$$\varrho_0 = p_0/gH, \qquad (5)$$

where $T_0 =$ basic temperature distribution,

$$H = RT_0/g, \qquad (6)$$

$$x = \int_0^z \frac{dz}{H}. \qquad (7)$$

We now investigate the equations based on the above approximations, and obtain solutions for the most common thermal tides and tides.* In the light of these solutions we examine the effects of approximations (g)–(j) referring, where possible, to earlier studies of these matters.

We use the equations in the form chosen by Siebert (1961, p. 128)† and Pekeris (1937); making much use of Equations (4) to (6):

$$\frac{\partial u}{\partial t} - 2\omega v \cos\theta = -\frac{1}{a}\frac{\partial}{\partial\theta}\left(\frac{\delta p}{\varrho_0} + \Omega\right), \qquad (8)$$

$$\frac{\partial v}{\partial t} + 2\omega u \cos\theta = -\frac{1}{a\sin\theta}\frac{\partial}{\partial\varphi}\left(\frac{\delta p}{\varrho_0} + \Omega\right), \qquad (9)$$

$$\frac{\partial\delta p}{\partial z} = -g\delta\varrho - \varrho_0\frac{\partial\Omega}{\partial z}, \qquad (10)$$

$$\frac{D\varrho}{Dt} \equiv \frac{\partial\delta\varrho}{\partial t} + w\frac{d\varrho_0}{dz} = -\varrho_0\chi, \qquad (11)$$

where

$$\chi \equiv \nabla\cdot\mathbf{V} = \frac{1}{a\sin\theta}\frac{\partial}{\partial\theta}(u\sin\theta) + \frac{1}{a\sin\theta}\frac{\partial v}{\partial\varphi} + \frac{\partial w}{\partial z}, \qquad (12)$$

$$\frac{R}{\gamma-1}\frac{DT}{Dt} \equiv \frac{R}{\gamma-1}\left(\frac{\partial\delta T}{\partial t} + w\frac{dT_0}{dz}\right) = \frac{gH}{\varrho_0}\frac{D\varrho}{Dt} + J, \qquad (13)$$

---

* Some of the approximations made above are avoided to some extent in a recent study by Hunt and Manabe (1968) using a numerical weather prediction model. However, that study introduced additional approximations (such as putting a top on the atmosphere near 38 km, and using 18 levels, equi-spaced in pressure to resolve the tidal vertical structure) which are seen in what follows to be more serious than the approximations used here. See also Lindzen, Batten and Kim (1968).

† But we use the gas constant R for air rather than the universal gas constant R_0 (cf. p. 3).

and

$$\delta p/p_0 = \delta T/T_0 + \delta\varrho/\varrho_0. \tag{14}$$

Here θ=colatitude; $\phi=$ east longitude; t=time; u=northerly velocity; v=westerly velocity; w=upward velocity; δp=pressure perturbation; $\delta\varrho$=density perturbation; δT=temperature perturbation; J=thermotidal heating per unit mass per unit time; Ω=gravitational tidal potential; ω=earth's rotation rate; $\gamma=C_p/C_v=1.4$, and for future reference $\kappa=(\gamma-1)/\gamma=2/7$.

Here (8) and (9) are the linearized equations for northerly and westerly momentum respectively (the coriolis terms in these equations describe the advection of the earth's momentum due to its rotation); (10) is the hydrostatic pressure relation; (11) is the continuity equation; (13) is the thermodynamic energy equation where heating is produced by both external sources (J) and 'adiabatic' compression; (14) is the linearized version of the perfect gas law. Using (14) to eliminate δT from (13), we obtain

$$\frac{Dp}{Dt} = \frac{\partial\delta p}{\partial t} + w\frac{dp_0}{dz} = \gamma g H\frac{D\varrho}{Dt} + (\gamma-1)\varrho_0 J. \tag{15}$$

In meteorological work Dp/Dt is often used as the main dependent variable (and usually denoted by ω); a related variable comparably convenient for tidal theory is

$$G = -\frac{1}{\gamma p_0}\frac{Dp}{Dt}, \tag{16}$$

which is used in our subsequent equations from (22) onwards.

In tidal theory we are generally concerned with fields periodic in time and longitude; i.e., of the form (using complex quantities)

$$f = f^{\sigma,s}(\theta, z)\, e^{i(\sigma t + s\phi)}, \tag{17}$$

where $2\pi/\sigma$ represents either a solar or lunar day or some suitable fraction thereof;

$$s = 0, \pm 1, \pm 2, \dots.$$

From (17) we have

$$\partial/\partial t = i\sigma; \quad \partial/\partial\phi = is.$$

Using (17) we may solve (8) and (9) for u and v:

$$u^{\sigma,s} = \frac{i\sigma}{4a\omega^2(f^2 - \cos^2\theta)}\left(\frac{\partial}{\partial\theta} + \frac{s\cot\theta}{f}\right)\left(\frac{\delta p^{\sigma,s}}{\varrho_0} + \Omega^{\sigma,s}\right), \tag{18}$$

and

$$v^{\sigma,s} = \frac{-\sigma}{4a\omega^2(f^2 - \cos^2\theta)}\left(\frac{\cos\theta}{f}\frac{\partial}{\partial\theta} + \frac{s}{\sin\theta}\right)\left(\frac{\delta p^{\sigma,s}}{\varrho_0} + \Omega^{\sigma,s}\right), \tag{19}$$

where

$$f \equiv \sigma/2\omega.$$

Here δp is a complex *function*, and

$$[\partial/\partial\theta + s\cot\theta/f] \quad \text{and} \quad [(\cos\theta/f)(\partial/\partial\theta) + s/\sin\theta]$$

are different operators; hence (18) and (19) do not imply that u lags $90°$ in phase behind v in the northern hemisphere. In addition, u and v may change sign at different latitudes; thus, there may exist a latitude band where u leads v (Blamont and Teitelbaum, 1968); (18) and (19) also appear to suggest that u and v become infinite when $f = \pm\cos\theta$. However, as Brillouin (1932) showed, this does not occur for complete solutions of our equations. Substituting (18) and (19) into (12) we get

$$\chi - \frac{\partial w}{\partial z} = \frac{i\sigma}{4a^2\omega^2}F\left(\frac{\delta p}{\varrho_0} + \Omega\right),\tag{20}$$

where

$$F \equiv \frac{1}{\sin\theta}\frac{\partial}{\partial\theta}\left(\frac{\sin\theta}{f^2-\cos^2\theta}\frac{\partial}{\partial\theta}\right) - \frac{1}{f^2-\cos^2\theta}\left(\frac{s}{f}\frac{f^2+\cos^2\theta}{f^2-\cos^2\theta} + \frac{s^2}{\sin^2\theta}\right).\tag{21}$$

Equations (20), (16), (15), (11) and (10) are five equations in five unknowns: G, δp, δp, w, and χ. They are readily reduced to a single equation for G alone:

$$H\frac{\partial^2 G^{\sigma,s}}{\partial z^2} + \left(\frac{dH}{dz} - 1\right)\frac{\partial G^{\sigma,s}}{\partial z} - \frac{i\sigma}{g}\frac{\partial^2\Omega^{\sigma,s}}{\partial z^2}$$
$$= \frac{g}{4a^2\omega^2}F\left(\left(\frac{dH}{dz} + \kappa\right)G^{\sigma,s} - \frac{\kappa J^{\sigma,s}}{\gamma g H}\right).\tag{22}$$

In general the scale length for the z-variation of Ω is the moon-earth or sun-earth distance. Consistently with the 'shallow atmosphere' approximation, therefore, $(\sigma/g)(\partial^2\Omega/\partial z^2)$ may be neglected, leaving us with

$$H\frac{\partial^2 G^{\sigma,s}}{\partial z^2} + \left(\frac{dH}{dz} - 1\right)\frac{\partial G^{\sigma,s}}{\partial z} = \frac{g}{4a^2\omega^2}F\left(\left(\frac{dH}{dz} + \kappa\right)G^{\sigma,s} - \frac{\kappa J^{\sigma,s}}{\gamma g H}\right).\tag{23}$$

This may be solved by the method of separation of variables. Assume that $G^{\sigma,s}$ may be written

$$G^{\sigma,s} = \sum_n L_n^{\sigma,s}(z)\Theta_n^{\sigma,s}(\theta)\tag{24}$$

and, moreover, that the set $\{\Theta_n^{\sigma,s}(\theta)\}_{\text{all }n}$ is complete for $0 \leqslant \theta \leqslant \pi$. Then J may be expanded thus:

$$J^{\sigma,s} = \sum_n J_n^{\sigma,s}(z)\Theta_n^{\sigma,s}(\theta).\tag{25}$$

Substitution of (24) and (25) into (23) yields the following set of equations for $L_n^{\sigma,s}$ and $\Theta_n^{\sigma,s}$:

$$F(\Theta_n^{\sigma,s}) = -\frac{4a^2\omega^2}{gh_n^{\sigma,s}}\Theta_n^{\sigma,s}\tag{26}$$

and

$$H \frac{d^2 L_n^{\sigma,s}}{dz^2} + \left(\frac{dH}{dz} - 1\right) \frac{dL_n^{\sigma,s}}{dz} + \frac{1}{h_n^{\sigma,s}} \left(\frac{dH}{dz} + \kappa\right) L_n^{\sigma,s} = \frac{\kappa}{\gamma g H h_n^{\sigma,s}} J_n^{\sigma,s} \qquad (27)$$

where $h_n^{\sigma,s}$ is the constant of separation.*

The boundary conditions on $\{\Theta_n\}$ are that they be bounded at the poles (i.e., at $\theta = 0, \pi$). With these conditions (26) defines an eigenfunction-eigenvalue problem where $\{h_n\}$ is the set of eigenvalues. Laplace (1799, 1825) first derived (26), which is therefore called *Laplace's Tidal Equation* for the free surface oscillations of a spherical ocean envelope. In his problem, however, h_n was replaced by h, the depth of the ocean; the eigenvalue was σ (which appears in the operator F). By historical analogy with this problem $\{h_n\}$ is often called the set of *equivalent depths* (Taylor, 1936). The eigenfunctions $\{\Theta_n\}$ are often called *Hough Functions* after Hough (1897, 1898), who pioneered in the solution of (26).

Equation (27) is an inhomogeneous equation which, given two boundary conditions, has a unique solution for the vertical structure of a given Hough mode (i.e., the complete field whose latitude variation is given by a Hough function); (27) is often called the *vertical structure equation*. The relevant boundary conditions are discussed at the end of this section. We follow the common practice of reducing (27) to canonical form by the change of variables given by (7), or (using (4))

$$x = -\log\left(\frac{p_0}{p_0(0)}\right)^{\dagger}, \qquad (28)$$

and

$$L_n = e^{x/2} y_n. \qquad (29)$$

Then (27) becomes

$$\frac{d^2 y_n}{dx^2} - \frac{1}{4}\left[1 - \frac{4}{h_n}\left(\kappa H + \frac{dH}{dx}\right)\right] y_n = \frac{\kappa J_n}{\gamma g h_n} e^{-x/2}. \qquad (30)$$

Equations (20), (16), (15), (14), (11) and (10) together with Equation (24) imply that

$$\delta p = \sum_n \delta p_n(x)\, \Theta_n, \qquad (31)$$

$$\delta \varrho = \sum_n \delta \varrho_n(x)\, \Theta_n, \qquad (32)$$

$$\delta T = \sum_n \delta T_n(x)\, \Theta_n, \qquad (33)$$

$$w = \sum_n w_n(x)\, \Theta_n, \qquad (34)$$

---

* The superscripts σ, s apply to all tidal fields, Hough functions, and equivalent depths. We will, therefore, generally omit them, their presence being understood.

† For those readers familiar with the meteorological use of pressure (rather than z) as a vertical coordinate, a somewhat simpler derivation of (30) is available wherein dp/dt rather than G is used as a basic variable (Nunn, 1967; Flattery, 1967), and pressure coordinates are used from the beginning.

where it may be shown that

$$\delta p_n = \frac{p_0(0)}{H(x)} \left[-\frac{\Omega_n(x)}{g} e^{-x} + \frac{\gamma h_n}{i\sigma} e^{-x/2} \left(\frac{dy_n}{dx} - \tfrac{1}{2} y_n \right) \right],\tag{35}$$

$$\delta \varrho_n = \frac{p_0(0)}{(gH)^2} \left\{ -\Omega_n e^{-x} \left(1 + \frac{1}{H}\frac{dH}{dx} \right) + \frac{\gamma g h_n}{i\sigma} e^{-x/2} \left[\left(1 + \frac{1}{H}\frac{dH}{dx} \right)\left(\frac{dy_n}{dx} - \frac{1}{2} \right) \right.\right.$$
$$\left.\left. + \frac{H}{h_n}\left(\kappa + \frac{1}{H}\frac{dH}{dx} \right) y_n \right] - \frac{\kappa J_n}{i\sigma} \right\},\tag{36}$$

$$\delta T_n = \frac{1}{R} \left\{ \frac{\Omega_n}{H}\frac{dH}{dx} - \frac{\gamma g h_n}{i\sigma} e^{x/2} \left[\frac{\kappa H}{h_n} + \frac{1}{H}\frac{dH}{dx_3}\left(\frac{d}{dx} + \frac{H}{h_n} - \frac{1}{2} \right) \right] y_n + \frac{\kappa J_n}{i\sigma} \right\},\tag{37}$$

$$w_n = -\frac{i\sigma}{g}\Omega_n + \gamma h_n e^{x/2}\left[\frac{dy_n}{dx} + \left(\frac{H}{h_n} - \frac{1}{2} \right) y_n \right].\tag{38}$$

From (18), (19) and (35) it also follows that

$$u = \sum_n u_n(x) U_n(\theta),\tag{39}$$

$$v = \sum_n v_n(x) V_n(\theta),\tag{40}$$

where

$$U_n = \frac{1}{f^2 - \cos^2\theta}\left(\frac{d}{d\theta} + \frac{s\cot\theta}{f} \right)\Theta_n,\tag{41}$$

$$V_n = \frac{1}{f^2 - \cos^2\theta}\left(\frac{\cos\theta}{f}\frac{d}{d\theta} + \frac{s}{\sin\theta} \right)\Theta_n,\tag{42}$$

$$u_n = \frac{\gamma g h_n e^{x/2}}{4a\omega^2}\left(\frac{dy_n}{dx} - \tfrac{1}{2} y_n \right),\tag{43}$$

$$v_n = \frac{i\gamma g h_n e^{x/2}}{4a\omega^2}\left(\frac{dy_n}{dx} - \frac{1}{2} y_n \right).\tag{44}$$

Here Ω has already been expanded thus:

$$\Omega = \sum_n \Omega_n(x) \Theta_n.\tag{45}$$

We now turn to the question of boundary conditions for (30) or (27). Given our assumption of a smooth spherical earth, the lower boundary condition is simply $w=0$ at $x=z=0$. (46)

Using (38) this implies that at $x=0$,

$$\frac{dy_n}{dx} + \left(\frac{H}{h_n} - \frac{1}{2} \right) y_n = \frac{i\sigma}{\gamma g h_n}\Omega_n.\tag{47}$$

For the upper boundary condition one generally requires that the kinetic energy density $\frac{1}{2}\varrho_0(z)\,\mathbf{v}\cdot\mathbf{v}$ shall remain bounded as $z\to\infty$. This, in turn, requires that $y_n(x)$ remains bounded as $x\to\infty$ (viz. Eckart, 1960; Wilkes, 1949). In some circumstances this condition is inadequate. For example, consider an atmosphere with an isothermal top where H is a constant and $J_n=0$; then (30) becomes

$$\frac{\mathrm{d}^2 y_n}{\mathrm{d}x^2} - \frac{1}{4}\left[1 - \frac{4\kappa H}{h_n}\right]y_n = 0. \tag{48}$$

For $h_n < 4\kappa H$ the solution of (48) is

$$y = Ae^{i\lambda x} + Be^{-i\lambda x}, \tag{49}$$

where

$$\lambda = \sqrt{\left[\frac{\kappa H}{h_n} - \frac{1}{4}\right]}. \tag{50}$$

Both terms in (49) are bounded (Wilkes, 1949). The term $e^{i\lambda x}$ is associated with upward propagation of energy (Wilkes, 1949) and $e^{-i\lambda x}$ with downward propagation. Thus $B=0$ if there are no energy sources at $x=\infty$. This result was also obtained by Eliassen and Palm (1961). It is generally referred to as the *radiation condition*. It may alternately be obtained by introducing a small damping (cf., Golitsyn, 1965; Booker and Bretherton, 1967; Giwa, 1967, among numerous others), or by considering the tidal problem from an initial-value point of view (Wurtele, 1953). We use the radiation condition in our calculations. However, as shown later, it is not always applicable. In particular, it is possible for viscosity and conductivity, both of which increase as $1/\varrho$, to cause reflections of tides at great altitudes if λ is very small.

3.3. Methods of Solution

In this section we indicate some common methods for solving (26) and (30), and outline the use of these solutions.

3.3A. LAPLACE'S TIDAL EQUATION

Equation (26) may be rewritten, as by Hough (1897, 1898), in the form

$$\frac{\mathrm{d}}{\mathrm{d}\mu}\left(\frac{1-\mu^2}{f^2-\mu^2}\frac{\mathrm{d}\Theta_n}{\mathrm{d}\mu}\right) - \frac{1}{f^2-\mu^2}\left[\frac{s}{f}\frac{f^2+\mu^2}{f^2-\mu^2} + \frac{s^2}{1-\mu^2}\right]\Theta_n + \frac{4a^2\omega^2}{gh_n}\Theta_n = 0, \tag{51}$$

where $\mu=\cos\theta$; this form has been studied for well over a century, and most current methods of solution are based on Hough's work. However, only in recent years have thorough investigations of its solutions and their properties been made, largely owing to the advent of the electronic digital computer. Such analyses have been made by Flattery (1967), Longuet-Higgins (1967), Dikii (1965, 1967), and Golitsyn and Dikii (1966). Lindzen (1967b) has followed a simpler, more approximate approach, to isolate the effects of a change of sign of the quantity $(f^2-\mu^2)$.

In this section we simply outline a formal solution procedure. Looking at (51), it is fairly clear that there are, in general, no 'obvious' closed solutions and that an expansion procedure is necessary.*

It turns out that (51) has regular solutions for the whole domain $0 \leqslant \theta \leqslant \pi$. Thus, we might, for example, seek power series solutions of (51). Unfortunately the substitution of such a series into (51) leads to an infinite set of fifth-order recursion relations, that is, each relation involves five of the power series coefficients. However, as Hough (1898) noted, the use of an expansion in associated Legendre Polynomials leads to a much more tractable set of third-order recursion relations. Briefly, let

$$\Theta_n^{\sigma, s} = \sum_{m=s}^{\infty} C_{n, m}^{\sigma, s} P_m^s(\mu). \tag{59}$$

Substitution of (59) into (51) leads, after considerable manipulation, to the following set of equations for $C_{n, m}$: at end

$$\frac{(m - s)}{(2m - 1)\left\{m(m - 1) - \dfrac{s}{f}\right\}}\left[\frac{(m - s - 1)}{(2m - 3)} C_{n, m-2} + \frac{(m - 1)^2(m + s)}{m^2(2m + 1)} C_{n, m}\right]$$

$$- \left[f^2 \frac{m(m + 1) - \dfrac{s}{f}}{m^2(m + 1)^2} - \frac{h_n g}{4\omega^2 a^2}\right] C_{n, m}$$

$$+ \frac{(m + s + 1)}{(2m + 3)\left\{(m + 1)(m + 2) - \dfrac{s}{f}\right\}}\left[\frac{(m + 2)^2(m - s + 1)}{(m + 1)^2(2m + 1)} C_{n, m}\right.$$

$$\left. + \frac{(m + s + 2)}{(2m + 5)} C_{n, m+2}\right] = 0, \tag{60}$$

---

* There are three exceptions:
  (a) When $s = 0, f = 1$, as Solberg (1936) first noticed, we can write
$$\Theta_n^{2\omega, 0} = \sin \tfrac{1}{2} n\pi\mu \quad n = 1, 3, 5,$$
$$\Theta_n^{2\omega, 0} = \cos \tfrac{1}{2} n\pi\mu \quad n = 0, 2, 4, \tag{52}$$
and then
$$4\omega^2 a^2/gh_n^{2\omega, 0} = (\tfrac{1}{2} n\pi)^2. \tag{53}$$

  (b) For a non-rotating earth, using the Neumann form (p. 37), we can write
$$\Theta_n^{\sigma, s} = N_n^s(\mu), \tag{54}$$
where N_n^s is the associated Legendre Polynomial as defined by Whittaker and Watson (1927); then
$$\sigma^2 = n(n + 1) gh_n^s/a^2. \tag{55}$$

  (c) when $h = \infty$,
$$\Theta_n^{\sigma, s} = R_n^s N_{n-1}^s + S_n^s N_{n+1}^s, \tag{56}$$
$$R_n^s = (n + 1), \sqrt{n + s}/n \tag{57a}$$
$$S_n^s = [n/(n + 1)] [(n + 1 - s)/\sqrt{n + s}], \tag{57b}$$
and
$$f = s/n(n + 1). \tag{58}$$

Flattery (1967) discussed these cases in some detail; see also Siebert (1961) for cases (a), (b), and Haurwitz (1940) and Neamtan (1946) for case (c).

Hough (1898) found it convenient to introduce a set of auxiliary constants $D_{n,m+1}$ and $D_{n,m}$ thus defined:

$$\frac{2(m+1)^2(m-s)}{m^2(2m-1)} C_{n,m-1} + \frac{(m+s+1)}{2m+3} C_{n,m+1}$$

$$= 2\left(\frac{m+1}{m} - \frac{s}{fm^2}\right) D_{n,m}. \quad (61)$$

It turns out that each $D_{n,m}$ can be interpreted as a coefficient in an expansion of the stream function for the horizontal flow (Love, 1913; Flattery, 1967; Longuet-Higgins, 1967). Substituting (61) into (60) we get

$$\frac{2(m+1)^2(m-s)}{m^2(2m-1)} D_{n,m-1} + \frac{(m+s+1)}{(2m+3)} D_{n,m+1}$$

$$= 2\left[f^2\left\{\frac{m+1}{m} - \frac{s}{fm^2}\right\} - (m+1)^2 \frac{h_n g}{4\omega^2 a^2}\right] C_{n,m}. \quad (62)$$

For brevity (61) and (62) may be rewritten thus:

$$K_m^s C_{n,m-1} - N_m^{\sigma,s} D_{n,m} + L_m^s C_{n,m+1} = 0, \quad (63)$$

and

$$K_m^s D_{n,m-1} - M_{n,m}^{\sigma,s} C_{n,m} + L_m^s D_{n,m+1} = 0, \quad (64)$$

where

$$K_m^s = \frac{2(m+1)^2(m-s)}{(2m-1)m^2} \quad (65)$$

$$L_m^s = \frac{(m+s+1)}{(2m+3)} \quad (66)$$

$$N_m^{\sigma,s} = 2\left\{\frac{m+1}{m} - \frac{1}{m^2}\frac{s}{f}\right\} \quad (67)$$

$$M_{n,m}^{\sigma,s} = 2\left[f^2\left\{\frac{m+1}{m} - \frac{1}{m^2}\frac{s}{f}\right\} - (m+1)^2 \frac{h_n^{\sigma,s} g}{4\omega^2 a^2}\right]. \quad (68)$$

Two things may be noted about (63) and (64): First, the equations for $\{C_{n,m}\}$, $(m-s)$ odd, $\{D_{n,m}\}$, $(m-s)$ even are decoupled from the equations for $\{C_{n,m}\}$, $(m-s)$ even, $\{D_{n,m}\}$, $(m-s)$ odd. The latter correspond to Hough Functions symmetric about the equator and the former to Hough Functions antisymmetric about the equator. Second, (63) and (64) are homogeneous. Thus, for (63) and (64) to have any solution, the coefficients must satisfy a self-consistency relation, and this relation yields the eigenvalues $\{h_n\}$; (63) and (64) are an infinite set of homogeneous linear equations for an infinite number of unknowns. By analogy with the case of N homogeneous equations in N unknowns, where self-consistency requires that the determinant of the coefficients equals zero, we require (ignoring certain mathematical technicalities) that the infinite determinant of the coefficients in (63) and (64) equals zero. For example; let σ (or f) and s be given, then our self-consistency relations are

$$\begin{vmatrix} -M_{n,s}^{\sigma,s} & L_s^s & 0 & 0 & 0 & \cdots \\ K_{s+1}^s & -N_{s+1}^{\sigma,s} & L_{s+1}^s & 0 & 0 & \cdots \\ 0 & K_{s+2} & -M_{n,s+2}^{\sigma,s} & L_{s+2}^s & 0 & \cdots \\ 0 & 0 & K_{s+3}^s & -N_{s+3}^{\sigma,s} & L_{s+3}^s & 0 \\ \vdots & \vdots & \vdots & \vdots & \vdots & \vdots \end{vmatrix} = 0 \tag{69}$$

(this yields the eigenvalues for symmetric eigenfunctions) and

$$\begin{vmatrix} -N_s^{\sigma,s} & L_s^s & 0 & 0 & 0 & \cdots \\ K_{s+1}^s & -M_{n,s+1}^{\sigma,s} & L_{s+1}^s & 0 & 0 & \cdots \\ 0 & K_{s+2}^s & -N_{s+2}^{\sigma,s} & L_{s+2}^s & 0 & \cdots \\ 0 & 0 & K_{s+3}^s & -M_{n,s+3}^{\sigma,s} & L_{s+3}^s & \cdots \\ \vdots & \vdots & \vdots & \vdots & \vdots & \vdots \end{vmatrix} = 0 \tag{70}$$

(which yields the eigenvalues for antisymmetric eigenfunctions). That (69) and (70), viewed as polynomials in h_n, have an infinite number of roots, merely corresponds to the infinity of values of n. Having obtained h_n from (69) or (70), one can obtain the eigenfunction by solving (63) and (64) for $\{C_{n,m}\}$ for all m. In general, if (59) converges, $C_{n,m} \xrightarrow[m \to \infty]{} 0$, and from (61) the same will hold for $D_{n,m}$. Hence a given solution of (69) or (70) is also an approximate solution of (69) or (70) when the determinants are truncated after a sufficiently large number of terms. Let $x = h_n g / 4\omega^2 a^2$ and let $D_l(x)$ be the $l \times l$ truncation of either (69) or (70). In practice we solve $D_l(x) = 0$ and see if the result is appreciably altered by considering $D_{l+1}(x) = 0$. We continue this process until the root is negligibly altered. This is essentially the method of Galerkin (Dikii, 1965); for an example, see Lindzen (1966a).

3.3B. VERTICAL STRUCTURE EQUATION

For reasonable choices of H, (30) is a well-behaved, non-singular differential equation. However, except for particularly simple choices of H, there are no simple closed form solutions. When H is a constant (isothermal atmosphere) or when $\kappa H + \mathrm{d}H/\mathrm{d}x$ is a constant (see Siebert (1961) for a discussion of this rather unrealistic case), (30) has homogeneous solutions which are either exponential or sinusoidal. When $\mathrm{d}H/\mathrm{d}z$ is constant, it may be shown that (30) has homogeneous solutions in the form of Bessel Functions. In either case Green's functions can be formed to handle the inhomogeneous problem. Treatments of tidal problems along the above lines have been given by Pekeris (1937), Siebert (1961), Butler and Small (1963), Lindzen (1967a), and others. For problems of any complexity, even the above methods usually require the use of a computer. This being the case, there are certain advantages in immediately approaching the solution of (30) numerically, as was done by Lindzen (1968a). The most obvious advantage of a numerical integration is the convenient flexibility it affords in the choice of H and J.

The rest of this section is devoted to a description of a particularly efficient numerical scheme, due originally to Bruce, Peaceman, Rachford and Rice (1953) and described by Richtmyer (1957). Let us divide our x-domain into a number of discrete levels $x_0, x_1, x_2 \dots$, where $x_0 = 0$ and the remaining levels are uniformly spaced, with separation δx. At x_m the derivatives of a function f may be approximated as follows:

$$\frac{\mathrm{d}f}{\mathrm{d}x} \approx \frac{(f)_{m+1} - (f)_{m-1}}{2\delta x} \tag{71}$$

$$\frac{\mathrm{d}^2 f}{\mathrm{d}x^2} \approx \frac{(f)_{m+1} - 2(f)_m + (f)_{m-1}}{(\delta x)^2}. \tag{72}$$

Here the subscripts refer to the level where f is evaluated. Using (71) and (72), a second order differential equation in f may be rewritten as

$$A_m (f)_{m+1} + B_m (f)_m + C_m (f)_{m-1} = D_m, \tag{73}$$

If we let $f = y_n(x)$, then from (30) we have

$$A_m = 1 \tag{74}$$

$$B_m = -\left[2 + \frac{(\delta x)^2}{4} \left\{ 1 - \frac{4}{h_n} \left(\kappa H(x_m) + \frac{\mathrm{d}H}{\mathrm{d}x} \bigg|_{x=x_m} \right) \right\} \right] \tag{75}$$

$$C_m = 1 \tag{76}$$

$$D_m = (\delta x)^2 \frac{\kappa J_n(x_m)}{\gamma g h_n}. \tag{77}$$

Our procedure in solving (73) is to let

$$(f)_m = \alpha_m (f)_{m+1} + \beta_m, \tag{78}$$

where α_m and β_m are new variables. Similarly

$$(f)_{m-1} = \alpha_{m-1} (f)_m + \beta_{m-1}. \tag{79}$$

Substituting (79) into (73) we get

$$(f)_m = -\frac{A_m (f)_{m+1}}{B_m + \alpha_{m-1} C_m} + \frac{D_m - \beta_{m-1} C_m}{B_m + \alpha_{m-1} C_m}. \tag{80}$$

Comparing (80) with (78) we obtain

$$\alpha_m = \frac{-A_m}{B_m + \alpha_{m-1} C_m} \tag{81}$$

$$\beta_m = \frac{D_m - \beta_{m-1} C_m}{B_m + \alpha_{m-1} C_m}. \tag{82}$$

From (81) and (82) we see that if we know α_0 and β_0 we can obtain all other α_m's and β_m's trivially; α_0 and β_0 are obtained from the condition $w = 0$ at $x = 0$. From (47) this implies that

$$\frac{(y_n)_1 - (y_n)_0}{\delta x} + \left(\frac{H_0}{h_n} - \frac{1}{2} \right) (y_n)_0 = \frac{i\sigma \Omega_{n,0}}{\gamma g h_n}; \tag{83}$$

this may be rewritten

$$(y_n)_0 = \frac{1}{1 - \left(\dfrac{H_0}{h_n} - \dfrac{1}{2}\right)} (y_n)_1 - \frac{i\sigma\Omega_{n,0}\,\delta x}{\gamma g h_n \left(1 - \left[\dfrac{H_0}{h_n} - \dfrac{1}{2}\right]\delta x\right)}. \tag{84}$$

Comparison of (84) with (78) yields

$$\alpha_0 = \frac{1}{1 - \left(\dfrac{H_0}{h_n} - \dfrac{1}{2}\right)\delta x} \tag{85}$$

$$\beta_0 = \frac{-i\sigma\Omega_{n,0}\,\delta x}{\gamma g h_n \left(1 - \left[\dfrac{H_0}{h_n} - \dfrac{1}{2}\right]\delta x\right)}. \tag{86}$$

We now merely have to know what y_n is at some high level, and (79) will give its value at all lower levels. The upper boundary condition permits us to do this. How it is done is most easily seen for an atmosphere for which dH/dx and J are zero above some level. From (49) and the discussion of p. 113, we have that

$$y_n = A e^{i\lambda x} \tag{87}$$

above this level; here A is some constant and

$$\lambda = \sqrt{\frac{\kappa H}{h_n} - \frac{1}{4}}.$$

From (87)

$$dy_n/dx = i\lambda y_n. \tag{88}$$

Let $m = M$ correspond to our top level. Applying (88) at level $M - 1$ we obtain

$$\frac{(y_n)_M - (y_n)_{M-2}}{2\delta x} = i\lambda (y_n)_{M-1}. \tag{89}$$

Now

$$(y_n)_{M-2} = \alpha_{M-2}(y_n)_{M-1} + \beta_{M-2} \tag{90}$$

and

$$(y_n)_{M-1} = \alpha_{M-1}(y_n)_M + \beta_{M-1}. \tag{91}$$

From (91), (90) and (89) we finally get

$$(y_n)_M = \frac{\beta_{M-2} + \beta_{M-1}(2i\lambda\delta x + \alpha_{M-2})}{(1 - \alpha_{M-1}[2i\lambda\delta x + \alpha_{M-2}])}. \tag{92}$$

Thus for an atmosphere for which dH/dx and J are zero above some level, our integration is completed. A similar method could be devised for other situations. There remains the technical matter of choosing δx. Referring to (30) we see that if the quantity λ^2 given by $\frac{1}{4}[4h_n^{-1}(\kappa H + dH/dx) - 1]$ is positive and 'sufficiently' slowly varying, we can interpret λ as a wave number and $2\pi/\lambda$ as a wavelength (in units of scale height). We have found that better than 1% accuracy for the integration is obtained by choosing

$$\delta x = (\text{minimum value of } 2\pi/\lambda(x)) \times 10^{-2}. \tag{93}$$

Similar considerations are appropriate to the case where $\lambda^2 < 0$. 'Slow' variation of λ^2 does appear to be a strong constraint.

The method of integration described above is extremely efficient despite the high resolution implied by (93). Dividing the interval $0 \leqslant z \leqslant 200$ km into 2000 subintervals (i.e., $\delta x \approx 0.1$ km/$H(x)$), we are able to integrate (30) on a CDC-6600 computer in less than one sec.

3.3.C. OUTLINE OF OVERALL PROCEDURE

In this section we show how the solutions of Equations (30) and (51) fit into a general computation of tidal fields. Let there be some excitation, J and/or Ω, with frequency σ and wave number s. We wish to compute the atmospheric response. Our first step is to find the roots of (69) and (70), thus obtaining $\{h_n\}_{\text{all } n}$ ('all n' is only in principle – in practice we settle for a finite number). Given this, we then solve (64) and (63) for $\{C_{n,m}\}_{\text{all } m}$. In general all $C_{n,m}, m > s$ (for even Hough functions) or $m > s+1$ (for odd Hough functions) are uniquely related to $C_{n,s}$ (or $C_{n,s+1}$). However, $C_{n,s}$ (or $C_{n,s+1}$) may be arbitrarily chosen. Thus the Hough functions obtained from (59) are determined only to within an arbitrary factor. We similarly use (41) and (42) to obtain $\{U_n(\mu)\}$ and $\{V_n(\mu)\}$ – also to within arbitrary factors.* The next is to expand $J^{\sigma,s}$ and/or $\Omega^{\sigma,s}$ in terms of $\{\Theta_n^{\sigma,s}\}$, and this requires a specification of the arbitrary factors. We now outline how this is done.

It may be shown (cf. Flattery, 1967) that the functions $\{\Theta_n\}_{\text{all } n}$ are orthogonal. Thus

$$\int\limits_{-1}^{1} \Theta_n(\mu) \Theta_{n'}(\mu) \, d\mu = (F_n)^2 \, \delta_{nn'}, \tag{95}$$

where

$$\delta_{nn'} = 0 \quad \text{if} \quad n \neq n'$$
$$= 1 \quad \text{if} \quad n = n',$$

and F_n is an undetermined factor.

If we set $C_{n,s}$ (or $C_{n,s+1}$) equal to one, then (63) and (64) determine the remaining $C_{n,m}$'s, and (59) gives Θ_n. Given the orthogonality of the associated Legendre Polynomials, we then have (in terms of the N_n^s of p.37)

$$(F_n)^2 = \int\limits_{-1}^{1} (\Theta_n(\mu))^2 \, d\mu = \sum_{m=s}^{\infty} (C_{n,m})^2 \int\limits_{-1}^{1} (N_m^s(\mu))^2 \, d\mu \tag{96}$$

* This step requires the use of the following property of $P_n{}^s$ (shared by $P_n{}^s$ and $P_{n,s}$, see pp. 35–37)

$$(1 - \mu^2) \frac{dP_n{}^s(\mu)}{d\mu} = -\frac{n(n - s + 1)}{2n + 1} P_{n+1}{}^s + \frac{(n + 1)(n + s)}{2n + 1} P_{n-1}{}^s. \tag{94}$$

(Whittaker and Watson, 1927). Because

$$\int_{-1}^{1} (N_m^s(\mu))^2 \, d\mu = \frac{2}{2m+1} \frac{(m+s)!}{(m-s)!},$$

(97)

(96) becomes

$$\int_{-1}^{1} (\Theta_n(\mu))^2 \, d\mu = \sum_{m=s}^{\infty} (C_{n,m})^2 \frac{2}{2m+1} \frac{(m+s)!}{(m-s)!}.$$

(98)

If normalized associated Legendre Polynomials ($P_{n,s}$ in the notation of p. 36) are used so that for $s>0$,

$$P_{m,s}(\mu) = \sqrt{\frac{2m+1}{2} \frac{(m-s)!}{(m+s)!}} N_m^s(\mu),$$

(99)

in which case

$$\Theta_n = \sum_{m=s}^{\infty} \hat{C}_{n,m}^{\sigma,s} P_{m,s}(\mu)$$

(100)

and

$$\hat{C}_{n,m} = C_{n,m} \sqrt{\frac{2}{2m+1} \frac{(m+s)!}{(m-s)!}}.$$

(101)

Similarly

$$(F_n)^2 = \sum_{m=s}^{\infty} (\hat{C}_{n,m}^{\sigma,s})^2.$$

(102)

If we choose to normalize Θ_n then we define

$$\bar{C}_{n,m} = \frac{\hat{C}_{n,m}}{F_n}$$

(103)

and

$$\bar{\Theta}_n = \sum_{m=s}^{\infty} \bar{C}_{n,m} P_{m,s}(\mu)$$

(104)

satisfies

$$\int_{-1}^{1} (\bar{\Theta}_n(\mu))^2 \, d\mu = 1.$$

(105)

From (102) and (103) we see that the normalization of Θ_n requires the knowledge of all the coefficients in the associated Legendre Polynomial expansion of Θ_n. Thus the truncation of the series in (102) will lead to small errors in $\bar{C}_{n,m}$. In practice this is not a serious problem, and we always use the normalized Hough Functions – omitting the overbar. If we expand some function $g(\mu)$ thus:

$$g(\mu) = \sum_n g_n \Theta_n(\mu),$$

(106)

then

$$g_n = \int_{-1}^{1} g(\mu)\, \Theta_n(\mu)\, d\mu \; ; \tag{107}$$

in particular, if

$$g(\mu) = P_{m,s}(\mu), \tag{108}$$

then

$$g_n = \bar{C}_{n,m}. \tag{109}$$

Having normalized the Hough Functions and expanded J and/or Ω, we can now solve (30) in order to obtain $\{y_n(x)\}$. Finally we use Equations (31)–(44) to evaluate u, v, w, δp, $\delta\varrho$, and δT.

3.4. Sources of Excitation

To apply the procedure described in Section 3.3.C. we need to know the excitations Ω and J. As already mentioned, these result from the rotation of the earth within gravitational tidal fields and radiational fields respectively. Tidal fields are determined by well known geometric considerations, thermal fields involve such uncertain factors as turbulence and atmospheric composition.

3.4A. GRAVITATIONAL EXCITATION

This is primarily due to the gravitational potential of the moon and secondarily to that of the sun. In analysing this excitation we follow the very lucid treatment of Lamb (1932). Consider Figure 3.1 where O and C denote the centers of the earth, and let $D = OC$, $a = OP$, $L = CP$, $\Theta = \angle POC$,

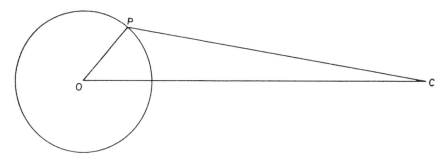

Fig. 3.1.   Geometry for calculation of tidal potentials; O is the earth's center, C is the center of the disturbing body (moon or sun). After Lamb (1932).

$\Delta = \angle NOC$, $\theta = \angle NOP =$ colatitude of P, where N denotes the north pole. The potential of the attraction of C at point P is $\gamma M/L$, where M denotes the mass of C and γ the gravitation constant. This may be rewritten

$$\Omega_{\text{local}} = \frac{\gamma M}{(D^2 - 2aD\cos\Theta + a^2)^{1/2}}. \tag{110}$$

It is the acceleration at P relative to the earth that produces tides. The potential associated with the acceleration of the earth as a whole is

$$\gamma \frac{M}{D^2} a \cos \Theta.$$

Subtracting this from (110) we get

$$\Omega_{\text{tidal}} = \frac{\gamma M}{(D^2 - 2aD \cos \Theta + a^2)^{1/2}} - \frac{\gamma M}{D^2} a \cos \Theta. \tag{111}$$

Expanding (111) in powers of (a/D), and retaining only the first term, we get

$$\Omega_{\text{tidal}} \approx -\frac{3}{2} \frac{\gamma M a^2}{D^3} (\tfrac{1}{3} - \cos^2 \Theta). \tag{112}$$

Let ϕ be the longitude of P, measured eastward from some fixed meridian, and let α be the hour angle of C west of the same meridian. Then

$$\cos \Theta = \cos \varDelta \cos \theta + \sin \varDelta \sin \theta \cos (\alpha + \phi), \tag{113}$$

and

$$\begin{aligned}
(\tfrac{1}{3} - \cos^2 \Theta) &= \tfrac{2}{3} (\cos^2 \varDelta - \tfrac{1}{3}) (\cos^2 \theta - \tfrac{1}{3}) \\
&\quad + \tfrac{1}{2} \sin 2\varDelta \sin 2\theta \cos (\alpha + \phi) \\
&\quad + \tfrac{1}{2} \sin^2 \varDelta \sin^2 \theta \cos 2(\alpha + \phi).
\end{aligned} \tag{114}$$

Now

$$\varDelta \approx \pi/2 + \varepsilon \sin \sigma_1 t \tag{115}$$

and

$$\alpha = \sigma_2 t. \tag{116}$$

If C is the moon then

$$\sigma_1^{\text{L}} = 2\pi/1 \text{ lunar month}, \tag{117}$$

and if C is the sun

$$\sigma_1^{\text{S}} = 2\pi/1 \text{ year}. \tag{118}$$

In both cases with superscript L or S to σ_2 and σ_1,

$$\sigma_2 = \frac{2\pi}{1 \text{ sidereal day}} - \sigma_1 = \sigma_0 - \sigma_1, \tag{119}$$

Let us consider the three terms on the right of (114) in order to see what frequencies they give rise to. We only make an approximate analysis, in which we assume that ε is small. Thus

$$\cos^2 \varDelta = \sin^2 (\varepsilon \sin \sigma_1 t) \approx \varepsilon^2 \sin^2 \sigma_1 t \approx \tfrac{1}{2} \varepsilon^2 (1 - \cos 2\sigma_1 t). \tag{120}$$

Since $(\cos^2 \Theta - \tfrac{1}{3})$ is independent of time, the first term in (114) is associated with either a lunar fortnightly or a solar semiannual excitation. Similarly

$$\sin 2\varDelta = + \sin (\pi + 2\varepsilon \sin \sigma_1 t) = - \sin (2\varepsilon \sin \sigma_1 t) \approx - 2\varepsilon \sin \sigma_1 t \tag{121}$$

and

$$\sin 2\Delta \cos(\alpha + \phi) \approx - \varepsilon(\sin(\alpha + \sigma_1 t + \phi) = \sin(\sigma_1 t - \alpha - \phi))$$

or using (116) and (119)

$$\sin 2\Delta \cos(\alpha + \phi) \approx - \varepsilon(\sin(\sigma_0 t + \phi) - \sin([\sigma_0 - 2\sigma_1]t + \phi)). \tag{122}$$

Thus the second term is associated with a sidereal diurnal excitation and an additional quasi-diurnal excitation with frequency $(\sigma_0 - 2\sigma_1)$.

Finally

$$\sin^2 \Delta = \cos^2(\varepsilon \sin \sigma_1 t) \approx 1 - \varepsilon^2 \sin^2 \sigma_1 t \approx \left(1 - \frac{\varepsilon^2}{2}\right) + \frac{\varepsilon^2}{2} \cos 2\sigma_1 t, \tag{123}$$

and

$$\sin^2 \Delta \cos 2(\alpha + \phi) \approx \left(1 - \frac{\varepsilon^2}{2}\right) \cos 2(\alpha + \phi)$$

$$+ \frac{\varepsilon^2}{4}\{\cos(2[\sigma_0 t + \phi]) + \cos(2([\sigma_0 - 2\sigma_1]t + \phi))\}. \tag{124}$$

Consequently the third term is associated with a relatively large lunar or solar semidiurnal term with frequency $2\sigma_2$, and much smaller terms associated with the frequencies $2\sigma_0$ and $2(\sigma_0 - 2\sigma_1)$.

A much more detailed and complete tidal analysis was given by Doodson (1922), and was later expounded by Bartels (1957); on this basis Siebert (1961) calculated the following contributions to Ω. (Note that $N_2^1 = \frac{3}{2} \sin 2\theta$, $N_2^2 = 3 \sin^2 \theta$.)

(diurnal lunar)

$$O_1 = - 6585. N_2^1(\theta) \sin[(\sigma_0 - 2\sigma_1^L)t + \phi]\,\mathrm{cm}^2/\mathrm{sec}^2 \tag{125}$$

(diurnal solar)

$$P_1 = - 3067. N_2^1(\theta) \sin[(\sigma_0 - 2\sigma_1^S)t + \phi]\,\mathrm{cm}^2/\mathrm{sec}^2 \tag{126}$$

(diurnal luni-solar)

$$K_{1m} + K_{1s} = + 9268. N_2^1(\theta) \sin(\sigma_0 t + \phi)\,\mathrm{cm}^2/\mathrm{sec}^2 \tag{127}$$

(large lunar elliptic semidiurnal)

$$N_2 = - 1518. N_2^2(\theta) \cos[(2\sigma_2^L - \sigma_1^S + v)t + 2\phi]\,\mathrm{cm}^2/\mathrm{sec}^2 \tag{128}$$

(semidiurnal lunar)

$$M_2 = - 7933. N_2^2(\theta) \cos[2(\sigma_2^L t + \phi)]\,\mathrm{cm}^2/\mathrm{sec}^2 \tag{129}$$

(semidiurnal solar)

$$S_2 = - 3700. N_2^2(\theta) \cos[2(\sigma_2^S t + \phi)]\,\mathrm{cm}^2/\mathrm{sec}^2 \tag{130}$$

(semidiurnal luni-solar)

$$K_{2m} + K_{2s} = - 1005. N_2^2(\theta) \cos[2(\sigma_0 t + \phi)]\,\mathrm{cm}^2/\mathrm{sec}^2; \tag{131}$$

(125)–(131) have been evaluated at $r = a$ (see p. 72 for M_2, N_2, O_1).

3.4.B. THERMAL EXCITATION DUE TO EXCHANGE OF HEAT WITH THE GROUND

Of all the solar radiation incident on the earth and its atmosphere system, most is absorbed by the ground and sea. The daily variations thus produced in the ground temperature are conveyed to the adjacent atmosphere by turbulence and infrared radiative transfer. Exactly how this occurs has been studied by Brunt (1939), Kuo (1968), Goody (1960), Kondratyev (1965), and others. In most of these studies the dynamic response of the atmosphere is neglected. In particular, the solutions obtained have temperature oscillations decaying with altitude. This heating can excite oscillations that the atmosphere will transmit upwards with amplitude growing as $1/\varrho^{1/2}$, but this is ignored. The method for estimating this omission was first developed by Chapman (1924a). One first calculates the transfer of the temperature oscillation, neglecting the large scale dynamic response. One then derives a heating function that would produce this oscillation. Finally, one uses this heating function in (13).

As an example of this approach let us consider the case where the surface temperature oscillation is transported to the atmosphere by turbulence, and where the turbulent transfer is modelled by eddy diffusion, the eddy K conductivity being taken to be constant; horizontal diffusion is neglected.

The temperature is then given by

$$\frac{\partial T}{\partial t} = K \frac{\partial^2 T}{\partial z^2}, \tag{132}$$

where

$$T = T_0^{\sigma,s}(\theta)\, e^{i(\sigma t + s\phi)} \quad \text{at} \quad z = 0. \tag{133}$$

Since (132) does not explicitly involve θ, we may expand T as

$$T^{\sigma,s} = \sum_n T_n^{\sigma,s}(z)\, \theta_n^{\sigma,s}(\theta), \tag{134}$$

where $T_n^{\sigma,s}(0)$ is obtained from (133). Equation (132) may be rewritten

$$i\sigma T_n^{\sigma,s} = K \frac{d^2 T_n^{\sigma,s}}{dz^2}, \tag{135}$$

which has the solution

$$T_n^{\sigma,s} = Ae^{-kz} + Be^{+kz}, \tag{136}$$

where

$$k = \sqrt{\frac{\sigma}{K}}\, e^{\phi i/4}. \tag{137}$$

If we assume that there is no heat flux from above, $B = 0$; $A = T_n(0)$. Thus

$$T_n = T_n(0)\, e^{-kz}. \tag{138}$$

If we wrote

$$\partial T_n/\partial t = J_n/c_p, \tag{139}$$

then the J_n that would give rise to (138) would be

$$J_n = i\sigma c_p \, T_n(0) \, e^{-kz} \, e^{i(\sigma t + s\phi)}; \tag{140}$$

(140) is then used in (27) to obtain the atmospheric tidal response to the surface heating.

There has been no rigorous justification of the above procedure. However, it would appear to be adequate for situations where J_n dominates the advective terms in (13) in the boundary layer: i.e., $0 \leqslant z < O(\sqrt{K/\sigma})$.

Chapman (1924a), Siebert (1961) and Kertz (1956a) have carried out calculations along the above lines. They found that surface temperature oscillations excite negligible migrating thermal tides in surface pressure in the absence of strong resonance magnification. Strong oscillations in the wind and temperature within the boundary layer, however, are generated.

3.4.C. THERMAL EXCITATION DUE TO DIRECT ATMOSPHERIC ABSORPTION OF INSOLATION

Although most insolation is absorbed at the ground, a significant amount is absorbed in the atmosphere by water vapor and ozone (and by O_2 to a lesser extent). The daily variation in heating due to this absorption is distributed throughout the bulk of the atmosphere, and is the most important of the tidal and thermotidal excitations.

Let us consider a gas, G, whose density distribution is $\varrho_G(\theta, \phi, z, t)$ and whose absorption coefficient as a function of wavelength is $K_G(\lambda)$.* Then J_G, the heating due to absorption by G, is given by

$$J_G(z, \theta, \phi, t) = \frac{\varrho_G}{\varrho} \int\limits_{\lambda} d\lambda \, K_G(\lambda) \, I_{0, \lambda} \exp\left(- K_G(\lambda) \int\limits_{s} \varrho_G \, ds'\right), \tag{141}$$

where the first integration is over all the absorption bands of G; $I_{0, \lambda}$ = intensity of radiation of wavelength λ incident on the atmosphere; S = distance from the sun of the point in question.

J_G of course, is non-zero only during the day, that is, when

$$0 \leqslant |\phi| \leqslant \cos^{-1}\left(- \frac{\tan \alpha}{\cot \theta}\right) = \phi_{max}. \tag{142}$$

Here α denotes the angle between the earth's axis and the normal to the plane of the ecliptic (Butler and Small, 1963). Now $I_{0, \lambda}$ is, by definition, a function of local time, i.e., $I_{0, \lambda} = I_{0, \lambda}(\sigma t + \phi)$ where $\sigma = 2\pi/1$ solar day. If, moreover, the time and longitude dependence of ϱ and ϱ_G are ignored,** then

$$J_G(z, \theta, \phi, t) = J_G(z, \theta, \sigma t + \phi).$$

This will be the case in all the explicit studies to be described in Section 3.5. However, in Section 3.6 the effects of relaxing this assumption will be briefly discussed. Given

---

* K_G may, in addition, depend on temperature and pressure.
** At least to the extent that they are not functions of local time.

(141) and (142) we may expand J_G as

$$J_G = \mathrm{Re} \sum_{n=0}^{\infty} J_G^{n\sigma,\,s}(z,\theta)\, e^{int'}, \tag{143}$$

where $t' = \sigma t + \phi$, and the $\{J_G^{n\sigma,\,s}(z,\theta)\}$ can be complex.

The term in (143) for which $n=0$ corresponds to the steady component of the heating. It is assumed that the atmosphere is in a state of quasi-equilibrium wherein there are constant (in time) cooling processes which exactly balance $J_G^{0,\,0}$.

The detailed evaluation of (141) when G is water vapor has been made by Mügge and Möller (1932) and Siebert (1961). The bands involved are primarily the near infrared bands at .94, 1.1, 1.38, 1.87, 2.7, 3.2 and 6.3 μ (Manabe and Möller, 1961). When G is O_3, calculations of (141) have been given by Pressman (1955), Johnson (1953), Butler and Small (1963) and Leovy (1964). The important absorption bands of O_3 are the Hartley and Huggins bands in the ultraviolet (2000–3700 Å) and the Chappuis band in the visible (4400–7600 Å). Absorption by CO_2 and O_2 appear to be significantly less important (Green, 1965). However, no thorough investigation of their role in exciting thermal tides has yet been made. Preliminary studies may be found in Manabe and Möller (1961), Leovy (1964) and Harris and Priester (1965). Here we restrict ourselves to excitation by H_2O and O_3 absorption.

It sometimes proves convenient to use instead of $J^{\sigma,\,s}$ a related function

$$\tau_G^{\sigma,\,s} = \frac{\kappa J_G^{\sigma,\,s}}{i\sigma R}. \tag{144}$$

By analogy with the procedure of Section 3.4.B this $\tau_G^{\sigma,\,s}$ corresponds, approximately, to the temperature oscillation that would be produced by $J_G^{\sigma,\,s}$ if the dynamic response of the atmosphere were ignored.

As Craig (1965) points out, the precise evaluation of (141) requires detailed knowledge of various factors like the distribution of ozone and water vapor and the solar ultraviolet spectrum – which are only roughly known at present. Thus the consideration of details of the heating and of the seasonal variations seems unwarranted at present, although some attempts to estimate seasonal effects have been made by Butler and Small (1963), Siebert (1961), Leovy (1964), and Lindzen (1967a). In view of this it is sufficient to consider J's which are separable in their latitude and altitude dependence; i.e.,

$$\tau = \sum_{G} \sum_{n=1}^{\infty} \tau_G^{n\sigma,\,n}(z,\theta)\, e^{int'}, \tag{145}$$

where

$$\tau_G^{\sigma,\,s} = f_G^{\sigma,\,s}(z)\, g_G^{\sigma,\,s}(\theta). \tag{146}$$

Figure 3.2 shows on the left $f_G(z)$ for both H_2O and O_3. The scale for $f_G(z)$ has been arbitrarily chosen, and differences between $f_G(z)$ for diurnal and semidiurnal excitations are ignored. The amplitudes of the τ_G's are incorporated into the $g_G(\theta)$'s, which are shown on the right side of Figure 3.2. Values corresponding to the diurnal and semidiurnal excitations ($n = 1$ and 2) are indicated. For $n = 1$ the phase corresponds to τ being a maximum at 1800 LT, while for $n = 2$ the phase corresponds to τ being a maximum at 0300 and 1500 LT. Differences in the shape of $g_G(\theta)$ for $n = 1$ and 2 have been ignored.

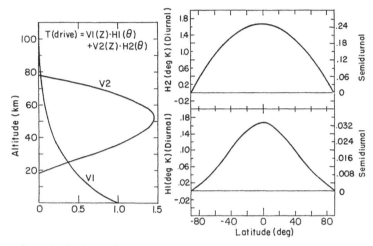

Fig. 3.2. Vertical distributions of thermal excitation due to water vapor (V1) and ozone (V2); latitude distributions for water vapor (H1) and ozone (H2). After Lindzen (1968a).

3.4.D. SUMMARY

In the preceding sections we have briefly described various sources of tidal and thermotidal excitation. It appears that insolation absorption by ozone and water vapor is the most important source of thermal excitation, and, in considering explicit solutions, we restrict our attention to these. The importance of heating due to the transfer of the daily variations of ground temperature to the atmosphere by means of turbulence and radiation has been estimated to be small – especially when averaged over both land and sea. The effects of daily variations in cloud cover and daily variations in the release of latent heat have not been evaluated. Indeed, our theoretical ignorance of turbulence and cloud processes makes our relegation of these processes to secondary importance somewhat questionable. Observational evidence is, however, more convincing. Heating due to surface temperature variations is markedly different over land and sea. As is shown in Section 3.6.A such differential heating will produce thermal tides that do not follow the sun. Haurwitz (1956, 1965) has analyzed surface pressure data from all over the world for the semidiurnal and diurnal oscillations. He found components that do not follow the sun (non-migrating); they are smaller than the migrating components. He did not consider thermotidal oscillations on a very small scale (geographically) including the sea breezes which are important non-migrating components, easily isolated.

Gravitational excitation is in general much weaker than thermal excitation. There is virtually no way of isolating solar gravitational tides from solar thermal tides in the data. Lunar tides, with their different period, are distinguishable. Despite their small

TABLE 3.1

Expansion coefficients relating normalized Hough functions $\Theta(L,M)$ and normalized associated Legendre functions $P(L,N)$ (the $P_{N,L}$ of p. 36). Symmetric gravitational modes (index M positive) with wave number $L = 2$ and period $S2 = 12.0000$ mean solar hours. Also shown are equivalent depths

	$\Theta(2,2)$	$\Theta(2,4)$	$\Theta(2,6)$	$\Theta(2,8)$	$\Theta(2,10)$	$\Theta(2,12)$	$\Theta(2,14)$	$\Theta(2,16)$	Sum of squares
$P(2, \ 2)$	0.969152	0.216046	0.093838	0.052364	-0.033432	-0.023225	-0.017096	-0.013128	0.999601
$P(2, \ 4)$	-0.245226	0.798445	0.421218	0.249960	-0.164044	-0.115651	-0.085888	-0.066330	0.989617
$P(2, \ 6)$	0.024633	-0.543993	0.464159	0.433414	-0.332801	-0.253228	-0.196413	-0.155882	0.937585
$P(2, \ 8)$	-0.001292	0.139613	-0.699495	0.077301	-0.279327	-0.299649	-0.272429	-0.236626	0.812785
$P(2, 10)$	0.000042	-0.019654	0.320141	-0.670489	0.241346	-0.038075	-0.161091	-0.204432	0.679873
$P(2, 12)$	-0.000001	0.001772	-0.079831	0.502316	0.469838	0.402107	0.188429	0.028361	0.677445
$P(2, 14)$	0.	-0.000112	0.012932	-0.191792	-0.612212	0.167588	0.376121	0.313968	0.679884
$P(2, 16)$	0.	0.000005	-0.001490	0.047158	0.338970	-0.596015	-0.133561	0.201921	0.530971
$P(2, 18)$	0.	0.	0.000129	-0.008274	-0.116833	0.479869	-0.443128	-0.335830	0.553137
$P(2, 20)$	0.	0.	-0.000009	0.001100	0.028510	-0.223420	0.562110	-0.192892	0.403906
$P(2, 22)$	0.	0.	0.000000	-0.000115	-0.005270	0.072124	-0.350749	0.543582	0.423735
$P(2, 24)$	0.	0.	0.	0.000010	0.000770	-0.017521	0.145846	-0.465676	0.238433
$P(2, 26)$	0.	0.	0.	-0.000001	-0.000092	0.003361	-0.044992	0.247028	0.063058
$P(2, 28)$	0.	0.	0.	0.	0.000009	-0.000526	0.010899	-0.094844	0.009114
$P(2, 30)$	0.	0.	0.	0.	-0.000001	0.000069	-0.002150	0.028300	0.000806
$P(2, 32)$	0.	0.	0.	0.	0.	-0.000008	0.000354	-0.006843	0.000047
$P(2, 34)$	0.	0.	0.	0.	0.	0.000001	-0.000050	0.001379	0.000002
$P(2, 36)$	0.	0.	0.	0.	0.	0.	0.000006	-0.000236	0.000000
$P(2, 38)$	0.	0.	0.	0.	0.	0.	-0.000001	0.000035	0.000000
$P(2, 40)$	0.	0.	0.	0.	0.	0.	0.	-0.000005	0.000000
$P(2, 42)$	0.	0.	0.	0.	0.	0.	0.	0.000001	0.000000
h (km)	7.8519	2.1098	0.9565	0.5425	0.3486	0.2427	0.1786	0.1368	

TABLE 3.2

Expansion coefficients relating normalized Hough functions $\Theta(L,M)$ and normalized associated Legendre functions $P(L,N)$. Anti-symmetric gravitational modes (index M positive) with wave number $L=2$ and period $S2 = 12.0000$ mean solar hours. Also shown are equivalent depths

	$\Theta(2,3)$	$\Theta(2,5)$	$\Theta(2,7)$	$\Theta(2,9)$	$\Theta(2,11)$	$\Theta(2,13)$	$\Theta(2,15)$	$\Theta(2,17)$	Sum of squares
$P(2,3)$	0.909763	0.342113	0.176363	-0.107229	-0.072075	-0.051814	-0.039085	-0.030568	0.997654
$P(2,5)$	-0.408934	0.645517	0.451203	-0.304461	-0.214983	-0.158845	-0.121861	-0.096373	0.975788
$P(2,7)$	0.071127	-0.643189	0.269489	-0.373180	-0.331281	-0.273489	-0.223615	-0.184135	0.899092
$P(2,9)$	-0.006673	0.225090	-0.708730	0.097325	-0.163311	-0.240947	-0.248974	-0.232919	0.763446
$P(2,11)$	0.000394	-0.043345	0.415768	0.588393	0.344898	0.083213	-0.068187	-0.142374	0.671746
$P(2,13)$	-0.000016	0.005401	-0.129704	-0.570466	0.325347	0.411524	0.267633	0.118861	0.703242
$P(2,15)$	0.000000	-0.000476	0.026244	0.263100	-0.621726	0.010186	0.302866	0.324230	0.653410
$P(2,17)$	0.	0.000031	-0.003788	-0.077111	0.413456	-0.535311	-0.251925	0.085554	0.534249
$P(2,19)$	0.	-0.000002	0.000413	0.016086	-0.166074	0.531455	-0.325999	-0.379661	0.560701
$P(2,21)$	0.	0.	-0.000035	-0.002544	0.046825	-0.286235	0.567087	-0.054384	0.408675
$P(2,23)$	0.	0.	0.000002	0.000318	-0.009976	0.105101	-0.412295	0.491154	0.422365
$P(2,25)$	0.	0.	0.	-0.000032	0.001680	-0.028866	0.193677	-0.505639	0.294018
$P(2,27)$	0.	0.	0.	0.000003	-0.000231	0.006247	-0.066773	0.303433	0.096569
$P(2,29)$	0.	0.	0.	-0.000000	0.000026	-0.001102	0.017997	-0.129440	0.017080
$P(2,31)$	0.	0.	0.	0.	-0.000003	0.000162	-0.003941	0.042594	0.001830
$P(2,33)$	0.	0.	0.	0.	0.000000	-0.000020	0.000720	-0.011319	0.000129
$P(2,35)$	0.	0.	0.	0.	0.	0.000002	-0.000112	0.002503	0.000006
$P(2,37)$	0.	0.	0.	0.	0.	0.	0.000015	-0.000470	0.000000
$P(2,39)$	0.	0.	0.	0.	0.	0.	-0.000002	0.000076	0.000000
$P(2,41)$	0.	0.	0.	0.	0.	0.	0.	-0.000011	0.000000
$P(2,43)$	0.	0.	0.	0.	0.	0.	0.	0.000001	0.000000
$P(2,45)$	0.	0.	0.	0.	0.	0.	0.	-0.000000	0.000000
h (km)	3.6665	1.3671	0.7062	0.4297	0.2885	0.2069	0.1556	0.1212	

amplitude, they are of great interest because their excitation is perfectly known. Thus we specifically describe the theory of the lunar semidiurnal tide.

3.5. Explicit Solutions

In this section we describe three examples of the calculations outlined in Sections 3.2 and 3.3: (a) The solar semidiurnal thermal tide, (b) The solar diurnal thermal tide, and (c) The lunar semidiurnal tide. The results explain many of the observed features described in Chapter 2.

3.5.A. THE MIGRATING SOLAR SEMIDIURNAL THERMAL TIDE

For the migrating solar semidiurnal tide, $s=2$ and $f=\sigma/2\omega=1$ (neglecting the difference between the solar and sidereal day). The solutions of (69) and (70) giving the equivalent depths $\{h_n\}$ and the expansion coefficients $\{C_{n,m}\}$ (or $\{\bar{C}_{n,m}\}$ as defined by (103)) have been known for some time (Wilkes, 1949; Kertz, 1957; Siebert, 1961). The most complete solution has been given by Flattery (1967). Following his notation, $n=2, 4, 6,...$ correspond to modes symmetric about the equator, and $n=3, 5, 7,...$ correspond to antisymmetric modes. Tables 3.1 and 3.2 (due to Flattery) give $\{h_n^{\sigma,s}\}$ and $\{\bar{C}_{n,m}^{\sigma,s}\}$ for symmetric and antisymmetric modes respectively.

In this review we deal only with the symmetric part of the tide. Figure 3.3 shows Θ_2, Θ_4, and Θ_6, and Figure 3.4 and 3.5 show the associated U and V functions on different scales.

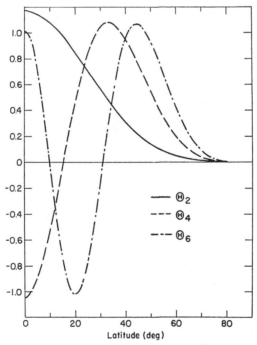

Fig. 3.3.   Latitude distribution for first three symmetric solar semidiurnal migrating Hough functions.

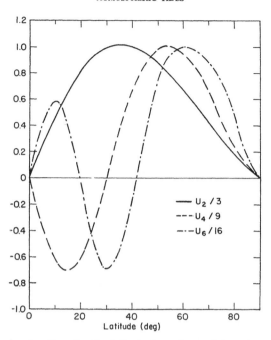

Fig. 3.4. The expansion functions for the latitude dependence of the solar semidiurnal component of u, the northerly velocity. The functions have been divided by the amounts shown.

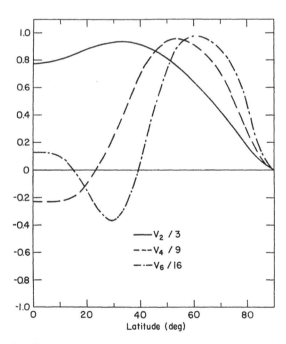

Fig. 3.5. The expansion functions for the latitude dependence of the solar semidiurnal component of v, the westerly velocity. The functions have been divided by the amounts shown.

Our next step is to expand g_{O_3} and g_{H_2O} in terms of the Hough Functions; the results (in degrees Kelvin) are:

$$g_{O_3}^{2\omega,\,2} = 0.249 \, K\Theta_2^{\sigma,\,s} + 0.0645 \, K\Theta_4^{\sigma,\,s} + 0.0365 \, K\Theta^{6,\,s} + \cdots \tag{147}$$

$$g_{H_2O}^{2\omega,\,2} = 0.0307 \, K\Theta_2^{\sigma,\,s} + 0.00796 \, K\Theta_4^{\sigma,\,s} + 0.00447 \, K\Theta_6^{\sigma,\,s} + \cdots . \tag{148}$$

A comparison of Figure 3.3 with Figure 3.2 shows that the latitude dependence of $\Theta_2^{2\omega,\,2}$ is quite similar to that of the g's; this corresponds to the dominance of the coefficients of $\Theta_2^{\sigma,\,s}$ in (147) and (148). The $\Theta_2^{2\omega,\,2}$ mode is associated with an equivalent depth of 7.852 km. For this equivalent depth, the quantity $\lambda^2 = \frac{1}{4}[4h^{-1}(\kappa H + dH/dx) - 1]$ in (30) is almost zero through most of the atmosphere – i.e.. the $\Theta_2^{2\omega,\,2}$ mode is associated with extremely long vertical wavelengths (ca. 150 km). Thus, not only does the $\Theta_2^{2\omega,\,2}$ mode receive the bulk of the semidiurnal excitation, but it must also respond to the excitation with particular efficiency – all the main excitations contribute 'in phase'. This immediately accounts for two of the most striking observed features of the migrating solar semidiurnal surface pressure oscillation (see Figure 1.4 and Section 2S.4A): namely, its strength and regularity; the latter results from the fact that the coefficients of $\Theta_2^{2\omega,\,2}$ in (147) and (148) depend on the overall latitude distribution of excitation, which does not change very much. The larger local variations in excitation will primarily affect the higher order, less efficient Hough modes.

If we confine our attention to the $\Theta_2^{2\omega,\,2}$ mode, and capitalize on the fact that $\lambda^2 \approx 0$, we may infer quite simply most of the significant features of the semidiurnal thermal tide (following Green, 1965). In this case (30) becomes

$$\frac{d^2 y_2^{2\omega,\,2}}{dx^2} \approx \frac{\kappa J_2^{2\omega,\,2}}{\gamma g h_2^{2\omega,\,2}} e^{-x/2} . \tag{149}$$

Let $J^{(i)}$ be the contribution to $J_2^{2\omega,\,2}$ from a very thin layer centered at $x = x^{(i)}$. Let $y^{(i)}$ be the specific response excited by $J^{(i)}$. Above $x^{(i)}$,

$$y^{(i)} = B_2^{(i)}, \tag{150}$$

and below $x^{(i)}$,

$$y^{(i)} = A_1^{(i)} x + B_1^{(i)}. \tag{151}$$

The lower boundary condition (47) implies that

$$A_1^{(i)} + \left(\frac{H}{h_2^{2\omega,\,2}} - \frac{1}{2} \right) B_1^{(i)} = 0,$$

or

$$B_1 \approx -2A_1 . \tag{152}$$

Thus, below $x^{(i)}$,

$$y^{(i)} = A_1^{(i)}(x - 2). \tag{153}$$

Near $x^{(i)}$, (149) has a particular solution for which

$$\frac{dy^{(i)}_{part}}{dx} \approx \frac{\kappa}{\gamma g h_2^{2\omega,\,2}} e^{-x^{(i)}/2} \int_{-\infty}^{\infty} J^{(i)} \, dx. \tag{154}$$

Integrating (154) we get

$$y^{(i)}_{part} \approx \frac{\kappa}{\gamma g h_2^{2\omega,\,2}} e^{-x^{(i)}/2} \left[\int_{-\infty}^{\infty} J^{(i)} \, dx \right] x. \tag{155}$$

Near $x^{(i)}$

$$y^{(i)} \approx y^{(i)}_{part} + A_1^{(i)} (x - 2). \tag{156}$$

A_1 and B_2 are determined by requiring that solutions for the various regions match:

$$A_1^{(i)} \approx -\frac{\kappa}{\gamma g h_2^{2\omega,\,2}} e^{-x^{(i)}/2} \left[\int_{-\infty}^{\infty} J^{(i)} \, dx \right], \tag{157}$$

and

$$B_2^{(i)} \approx -\frac{\kappa}{\gamma g h_2^{2\omega,\,2}} e^{-x^{(i)}/2} \left[\int_{-\infty}^{\infty} J^{(i)} dx \right] (x^{(i)} - 2); \tag{158}$$

(158) immediately tells us that *above the major sources of excitation*, the contributions from excitations below $x=2$ will be 180° out of phase with the contributions from excitations above $x=2$. This inference by Green (1965) has been completely confirmed by the more detailed calculations of Lindzen (1968a). We now use the above analysis to calculate the surface pressure oscillation. From (35)

$$\frac{\delta p_2^{2\omega,\,2}}{p_0} = \frac{1}{H} \frac{\gamma h_2^{2\omega,\,2}}{i\sigma} e^{x/2} \left(\frac{d}{dx} - \frac{1}{2} \right) y_2^{2\omega,\,2} \approx \frac{\gamma}{i\sigma} e^{x/2} \left(\frac{d}{dx} - \frac{1}{2} \right) y_2^{2\omega,\,2}. \tag{159}$$

For $x < x^{(i)}$ we have from (159), (157) and (158) that

$$\frac{\delta p^{(i)}}{p_0} \approx -\frac{1}{2} \frac{\kappa}{i\sigma g h_2^{2\omega,\,2}} e^{(x - x^{(i)}/2)} \left[\int_{-\infty}^{\infty} J^{(i)} \, dx \right] (x - 4). \tag{160)*}$$

At the ground (using $h_2^{2\omega,\,2} \approx H$)

$$\frac{\delta p^{(i)}}{p_0} \approx -\frac{2\kappa}{i\sigma g H} e^{-x^{(i)}/2} \left[\int_{-\infty}^{\infty} J^{(i)} \, dx \right],$$

---

* An implication of Equation (160) is that if the main excitation comes from above $x = 4$ – as appears to be the case – then the phases of the horizontal wind components will undergo 180° shifts at $x = 4$ (Green, 1965; Wilkes, 1949; Butler and Small, 1963). As yet, there has been no convincing observational test of this prediction.

or using (144),

$$\frac{\delta p^{(i)}}{p_0} \approx -\frac{2R}{gH} e^{-x^{(i)}/2} \left[\int_{-\infty}^{\infty} \tau^{(i)} \frac{dz}{H} \right].$$ (161)

Approximating the integral in (161) we have

$$\frac{\delta p^{(i)}}{p_0} \approx -\frac{2R}{gH^2} e^{-(x^{(i)}/2)} \Delta z_i \tau^{(i)},$$ (162)

where $\Delta z_i \approx$ thickness of ith layer.
Using $H = 7.6$ km we have

$$\frac{2R}{gH^2} \approx 1.014 \times 10^{-3} \, \text{deg}^{-1} \, \text{km}^{-1}.$$ (163)

Let $i = 1$ correspond to water vapor excitation. From Figure 3.2 we have

$$
\begin{aligned}
x^{(i)} &\approx 1.5, \\
e^{-x^{(i)}/2} &\approx .472, \\
\Delta z^{(i)} &\approx 18 \, \text{km}, \\
\tau^{(i)} &\approx 0.035 \, \text{K}.
\end{aligned}
$$ (164)

From (162) and (163) we then get

$$\delta p^{(1)} \approx 0.302 \times 10^{-3} \, p_0.$$ (165)

Let $i = 2$ correspond to ozone excitation. From Figure 3.2 we have

$$
\begin{aligned}
x^{(2)} &\approx 6, \\
e^{-x^{(i)}/2} &\approx 0.042, \\
\Delta_z^{(2)} &\approx 40 \, \text{km}, \\
\tau^{(2)} &\approx .5 \, \text{K}.
\end{aligned}
$$ (166)

From (162) and (163) we get

$$\delta p^{(2)} \approx 0.85 \times 10^{-3} \, p_0.$$ (167)

(165) and (167) are in excellent agreement with the more precise calculations of Siebert (1961), Butler and Small (1963) and Lindzen (1968a); (165) and (167) show that ozone is considerably more important than water vapor in exciting semidiurnal oscillations. This is because ozone excitation occurs over a greater depth than water vapor excitation, and at higher altitudes.

In following Green's treatment we have ignored the variation of H with altitude. Indeed, it has been shown by Lindzen (1968a) that the surface pressure oscillation is not sensitive to temperature structure. There are, of course, various unrealistic temperature structures for which this is not true. In particular, Siebert (1961) considered an atmosphere which was so cold above the tropopause that λ^2 became negative. In this atmosphere, oscillations excited by ozone could not propagate to the ground.

While the surface pressure oscillation is not particularly sensitive to temperature structure, the semidiurnal thermotidal oscillations in the ionosphere are. This is because below the mesopause, H is small and dH/dx is negative. Hence λ^2 is negative for the mode associated with $h_2^{2,2}$, and energy in this mode is trapped. Lindzen (1968a) studied the solar semidiurnal thermal tide for the equinoctial excitations shown in Figure 3.2. The procedure used was that outlined in Section 3.3. Only three Hough modes were used: $\Theta_2^{2\omega,2}$, $\Theta_4^{2\omega,2}$ and $\Theta_6^{2\omega,2}$. Three common temperature profiles were used – namely, the ARDC standard, the equatorial standard and an isothermal atmosphere at 260K. These are shown in Figure 3.6. The altitude distribution of the amplitudes of the resulting semidiurnal zonal wind oscillations over the equator are shown in Figure 3.7. Note that the oscillation in the ionosphere is smallest for the ARDC profile, which has the coldest mesopause. The diminution of growth below the mesopause is slightly less than we might expect from the behavior of the $\Theta_2^{2\omega,2}$ mode alone, since the higher modes are not significantly attenuated. For the most extreme case, the ARDC profile, the ratio $/y_2^{2\omega,2}/ : /y_4^{2\omega,2}/ : /y_6^{2\omega,2}/$ is $1:1.5\times10^{-2}:$ $8.\times10^{-3}$ at the ground, and $1:.3:.1$ at 100 km. It is clear that given a modest distortion in the latitude distribution of excitation, the $y_4^{2\omega,2}$ mode might play a significantly greater role in upper atmosphere fields.

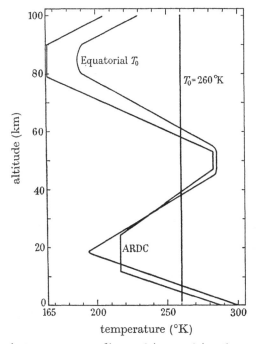

Fig. 3.6.   Different basic temperature profiles used in examining the semidiurnal thermal tide. After Lindzen (1968a); Minzner, Champion and Pond (1959).

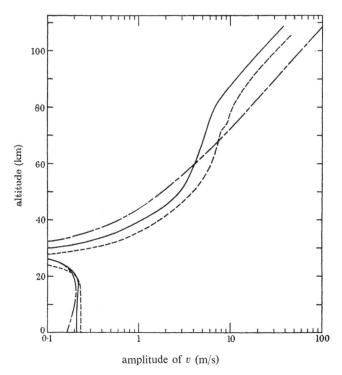

Fig. 3.7. Amplitude of the solar semidiurnal component of v over the equator for different basic temperature profiles; ———, ARDC; ---, equatorial; ——— – ———, isothermal ($T_0 = 260\,\mathrm{K}$). After Lindzen (1968a).

For a detailed description of the theoretical predictions of other semidiurnal fields at other latitudes the reader is referred to Nunn (1967). Given our uncertainties over the appropriate thermal structure, excitation and the effects of various excitations, such predictions must be taken as suggestive rather than definitive. Some idea of what is expected may be seen in Figures 3.8 and 3.9, where the altitude distribution of the amplitude and phase of the migrating solar semidiurnal thermal tide's northerly velocity component at various latitudes are shown. For T_0 we have taken the equatorial profile of Figure 3.6. The excitations are those shown in Figure 3.2. The computational procedure was that described in Section 3.3. Particularly noteworthy in Figures 3.8 and 3.9, are the relatively large changes in amplitude and phase with height in the ionosphere.

Before ending this section, we devote a few words to the resonance theory of the solar semidiurnal tide. It should be clear to the reader that given the presently known sources of thermal excitation, the amplitude of the migrating solar semidiurnal surface pressure oscillation is no surprise – nor does it depend on the atmosphere being highly tuned. However, it is only recently that we have come to know these sources of excitation. Previously it was believed that the observed oscillation was excited by a combination of gravitational drive and thermal exchange with the ground

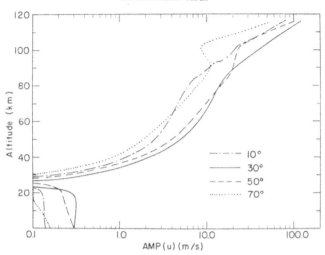

Fig. 3.8.   Amplitude of the solar semidiurnal component of u at various latitudes;
equatorial $T_0(z)$ assumed.

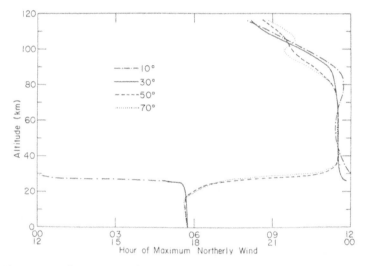

Fig. 3.9.   Phase (hour of maximum) of the solar semidiurnal component of u at various latitudes;
equatorial $T_0(z)$ assumed.

(Chapman, 1924a). Given this belief. the observed amplitude was, indeed, surprising and suggested a substantial atmospheric amplification (viz. Chapter 1). This would occur if the $\Theta_2^{2\omega, 2}$ mode were very close to a free oscillation of the atmosphere – i.e., an oscillation that could exist in the absence of continued excitation. Now if $J_n^{\sigma, s} = \Omega_n^{\sigma, s} = 0$, the solution to (30) is in general zero. However, for the atmosphere's thermal structure, as we presently know it, there exists one value of h for which (30) has a homogeneous solution – namely $h \approx 10.5$ km (Dikii, 1965). This is sufficiently far from $h = 7.852$ km to preclude a resonance. At the time the resonance theory was developed the thermal structure of the upper atmosphere was not known. Thus Taylor (1936) found that for an atmosphere where dT/dz was constant, there would be an infinite number of h's for which (30) would have homogeneous solutions. Pekeris (1937) found that more 'realistic' thermal structures might still have more than one such h; in particular, he found that for an atmosphere with a mesopeak temperature of about 350 K – which we now know to be too high – one of these h's would be near 7.852 km. For an extensive review of the resonance theory, the reader is referred to the monograph by Wilkes (1949).

3.5.B. THE SOLAR DIURNAL THERMAL TIDE

For this thermal tide $s=1$ and $f=\sigma/2\omega=\frac{1}{2}$ (again neglecting the difference between the solar and sidereal day). The complete solutions of (69) and (70) for these parameters have only recently been obtained (Lindzen, 1966a; Kato, 1966; Flattery, 1967). The most extensive solutions are again to be found in Flattery (1967).

Tables 3.3 and 3.4, taken from his work, show $\{h_n^{\omega,\,1}\}$ and $\{\bar{C}_{n,\,m}^{\omega,\,1}\}$ for symmetric modes. Tables 3.5 and 3.6 give these for antisymmetric modes. The reader will notice that we now have both positive and negative equivalent depths. Following Flattery's notation, $n=1, 3, 5...$ and $n=-2, -4...$ correspond to symmetric modes; $n=2, 4, 6...$ and $n=-1, -3, -5...$ correspond to antisymmetric modes. The negative indices are used for modes with negative equivalent depths. $\Theta_{-2}^{\omega,\,1}$, $\Theta_{-4}^{\omega,\,1}$, $\Theta_{1}^{\omega,\,1}$, $\Theta_{3}^{\omega,\,1}$, and $\Theta_{5}^{\omega,\,1}$ are shown in Figure 3.10. The corresponding U and V functions are shown in Figures 3.11 and 3.12.

The existence of negative equivalent depths may, at first, seem puzzling. However, their existence is readily understood in terms of the theory of internal gravity waves on a rotating plane where the rotation vector and gravity are parallel (viz. Eckart, 1960) – atmospheric tides may, in fact, be looked upon as global internal gravity waves resulting from a particular excitation. We know that internal gravity waves will propagate vertically in a rotating, planar fluid only if their period is greater than the Brunt-Väisälä period (ca. 5 min) and less than (twice the rotation rate 2π). Now the earth is not a plane, and twice the vertical component of its rotation rate varies from zero at the equator to $\pm 2\,\omega$ at the poles. At $\pm 30°$ latitude, twice the rotation rate equals ω. Extrapolating from the results for a rotating plane we would expect diurnal oscillations to propagate vertically equatorwards of $\pm 30°$, and to be trapped near the levels of excitation polewards of $\pm 30°$. If we look at Equation (30) we see that negative h's are associated with vertical trapping while sufficiently small positive h's are associated with vertical propagation. Reference to Figure 3.10 shows that the Hough Functions associated with negative h's have their largest amplitudes equatorwards of $\pm 30°$. Thus our intuitive extrapolation of gravity wave theory is shown to be approximately correct.* The above reasoning also explains the absence of negative h's for $f=1$ (i.e., semidiurnal tides). Finally it is important to notice the size of the positive h's: they are very small. Hence λ^2 will in general be large, and this,

---

* It appears that the high latitude trapping mechanism described here is also basic to the trapping of large scale, long period waves as described by Charney and Drazin (1961).

TABLE 3.3

Expansion coefficients relating normalized Hough functions $\Theta(L,M)$ and normalized associated Legendre functions $P(L,N)$ (the $P_{N,L}$ of p. 36). Symmetric gravitational modes (index M positive) with wave number $L = 1$ and period $S1 = 24.0000$ mean solar hours. Also shown are equivalent depths

	$\Theta(1,1)$	$\Theta(1,3)$	$\Theta(1,5)$	$\Theta(1,7)$	$\Theta(1,9)$	$\Theta(1,11)$	$\Theta(1,13)$	$\Theta(1,15)$	Sum of squares
$P(1,\ 1)$	0.282710	−0.073423	0.036899	0.023099	−0.016181	0.012140	0.009541	−0.007753	0.087771
$P(1,\ 3)$	−0.638229	0.156872	−0.078275	−0.048891	0.034214	−0.025657	−0.020157	0.016377	0.442966
$P(1,\ 5)$	0.620521	−0.060653	0.024423	0.014115	−0.009529	0.007009	0.005443	−0.004389	0.389710
$P(1,\ 7)$	−0.336408	−0.236676	0.131417	0.084483	−0.059835	0.045146	0.035596	−0.028986	0.201319
$P(1,\ 9)$	0.117021	0.512074	−0.237908	−0.144105	0.099367	−0.073914	−0.057788	0.046802	0.374147
$P(1,11)$	−0.028332	−0.580640	0.152307	0.068801	−0.040206	0.027050	0.019824	−0.015368	0.368854
$P(1,13)$	0.005042	0.459878	0.103730	0.109830	−0.088515	0.070577	0.057295	−0.047477	0.252689
$P(1,15)$	−0.000686	−0.278898	−0.374332	−0.248276	0.171313	−0.126402	−0.098089	0.078982	0.340712
$P(1,17)$	0.000074	0.135552	0.517192	0.221385	−0.115614	0.070486	0.047770	−0.034847	0.356704
$P(1,19)$	−0.000006	−0.054346	−0.499883	−0.026381	−0.056172	0.067486	0.063610	−0.056792	0.268514
$P(1,21)$	0.000000	0.018333	0.381800	−0.229616	0.221997	−0.176040	−0.139327	0.112718	0.311221
$P(1,23)$	0.	−0.005283	−0.242494	0.422033	−0.259734	0.157575	0.102752	−0.071711	0.344935
$P(1,25)$	0.	0.001316	0.131728	−0.487882	0.135377	−0.011050	0.030356	−0.043317	0.276629
$P(1,27)$	0.	−0.000286	−0.062318	0.439994	0.087035	−0.168519	−0.161015	0.139425	0.278817
$P(1,29)$	0.	0.000055	0.026007	−0.332074	−0.302585	0.263504	0.186488	−0.131860	0.324106
$P(1,31)$	0.	−0.000009	−0.009667	0.217009	0.432306	−0.213137	−0.079205	0.015959	0.286031
$P(1,33)$	0.	0.000001	0.003226	−0.125291	−0.454410	0.041738	−0.096555	0.128007	0.249648
$P(1,35)$	0.	−0.000000	−0.000972	0.064771	0.394455	0.170811	0.234046	−0.196474	0.282386
$P(1,37)$	0.	0.	0.000266	−0.030273	−0.296349	−0.342282	−0.254071	0.137926	0.289472
$P(1,39)$	0.	0.	−0.000067	0.012887	0.197576	0.426801	0.145742	0.016157	0.242863
$P(1,41)$	0.	0.	0.000015	−0.005026	−0.118725	−0.422911	0.040247	−0.177029	0.225934
$P(1,43)$	0.	0.	−0.000003	0.001804	0.064990	0.358427	−0.229362	0.257043	0.251375
$P(1,45)$	0.	0.	0.000001	−0.000599	−0.032662	−0.268838	0.362232	−0.216677	0.251502
$P(1,47)$	0.	0.	−0.000000	0.000184	0.015162	0.181957	−0.414474	0.075980	0.210900
$P(1,49)$	0.	0.	0.000000	−0.000053	−0.006532	−0.112537	0.394542	0.106833	0.179784
$P(1,51)$	0.	0.	0.	0.000014	0.002622	0.064165	−0.328983	−0.269253	0.184851
$P(1,53)$	0.	0.	0.	−0.000004	−0.000984	−0.033949	0.246673	0.370348	0.199159
$P(1,55)$	0.	0.	0.	0.000001	0.000346	0.016754	−0.168954	−0.399320	0.188282
$P(1,57)$	0.	0.	0.	−0.000000	−0.000155	−0.007743	0.106823	0.369244	0.147812
$P(1,59)$	0.	0.	0.	0.000000	0.000036	0.003363	−0.062816	−0.304303	0.096557

Table 3.3 (continued)

	$\Theta(1,1)$	$\Theta(1,3)$	$\Theta(1,5)$	$\Theta(1,7)$	$\Theta(1,9)$	$\Theta(1,11)$	$\Theta(1,13)$	$\Theta(1,15)$	Sum of squares
$P(1,61)$	0.	0.	0.	0.	-0.000011	-0.001377	0.034550	0.228240	0.053289
$P(1,63)$	0.	0.	0.	0.	0.000003	0.000532	-0.017854	-0.157853	0.025236
$P(1,65)$	0.	0.	0.	0.	-0.000001	-0.000195	0.008699	0.101570	0.010392
$P(1,67)$	0.	0.	0.	0.	0.000000	0.000068	-0.004008	-0.061202	0.003762
$P(1,69)$	0.	0.	0.	0.	-0.000000	-0.000022	0.001751	0.034706	0.001208
$P(1,71)$	0.	0.	0.	0.	0.	0.000007	-0.000727	-0.018595	0.000346
$P(1,73)$	0.	0.	0.	0.	0.	-0.000002	0.000287	0.009444	0.000089
$P(1,75)$	0.	0.	0.	0.	0.	0.000001	-0.000108	-0.004558	0.000021
$P(1,77)$	0.	0.	0.	0.	0.	-0.000000	0.000039	0.002096	0.000004
$P(1,79)$	0.	0.	0.	0.	0.	0.000000	-0.000013	-0.000920	0.000001
$P(1,81)$	0.	0.	0.	0.	0.	-0.000000	0.000004	0.000386	0.000000
$P(1,83)$	0.	0.	0.	0.	0.	0.	-0.000001	-0.000155	0.000000
$P(1,85)$	0.	0.	0.	0.	0.	0.	0.000000	0.000060	0.000000
$P(1,87)$	0.	0.	0.	0.	0.	0.	-0.000000	-0.000022	0.000000
$P(1,89)$	0.	0.	0.	0.	0.	0.	0.000000	0.000008	0.000000
$P(1,91)$	0.	0.	0.	0.	0.	0.	0.	-0.000003	0.000000
$P(1,93)$	0.	0.	0.	0.	0.	0.	0.	0.000001	0.000000
$P(1,95)$	0.	0.	0.	0.	0.	0.	0.	-0.000000	0.000000
$P(1,97)$	0.	0.	0.	0.	0.	0.	0.	0.000000	0.000000
h (km)	0.6909	0.1203	0.0484	0.0260	0.0162	0.0110	0.0080	0.0061	

TABLE 3.4

Expansion coefficients relating normalized Hough functions $\Theta(L,M)$ and normalized associated Legendre functions $P(L,N)$. Symmetric rotational modes (index M negative) with wave number $L = 1$ and period $S1 = 24.0000$ mean solar hours. Also shown are equivalent depths.

	$\Theta(1,-2)$	$\Theta(1,-4)$	$\Theta(1,-6)$	$\Theta(1,-8)$	$\Theta(1,-10)$	$\Theta(1,-12)$	$\Theta(1,-14)$	$\Theta(1,-16)$	Sum of squares
$P(1, 1)$	0.896764	-0.270454	-0.135299	0.083446	-0.057710	-0.042876	0.033445	-0.027020	0.909617
$P(1, 3)$	0.440182	0.470111	0.267139	-0.170324	0.119385	0.089298	-0.069924	0.056628	0.545460
$P(1, 5)$	0.045288	0.771088	0.111506	-0.014242	-0.006343	-0.010911	0.011317	-0.010574	0.609663
$P(1, 7)$	0.002067	0.326306	-0.615773	0.383244	-0.252535	-0.180576	0.137124	-0.108717	0.759536
$P(1, 9)$	0.000054	0.068707	-0.647410	-0.080047	0.185537	0.181976	-0.160164	0.138081	0.542526
$P(1, 11)$	0.000001	0.008838	-0.303072	-0.619316	0.339049	0.170143	-0.086221	0.043492	0.628712
$P(1, 13)$	0.	0.000773	-0.086241	-0.576203	-0.196000	-0.303630	0.273542	-0.226785	0.596311
$P(1, 15)$	0.	0.000049	-0.016874	-0.291010	-0.598964	-0.259151	0.044633	0.051946	0.515579
$P(1, 17)$	0.	0.000002	-0.002434	-0.098362	-0.528928	0.264995	-0.353422	0.280527	0.563271
$P(1, 19)$	0.	0.	-0.000271	-0.024366	-0.282509	0.575080	-0.180689	-0.069909	0.448658
$P(1, 21)$	0.	0.	-0.000024	-0.004664	-0.107059	0.494614	0.306951	-0.362368	0.481656
$P(1, 23)$	0.	0.	-0.000002	-0.000715	-0.030973	0.275870	0.552262	-0.112796	0.394780
$P(1, 25)$	0.	0.	-0.000000	-0.000090	-0.007145	0.133620	0.468226	0.332884	0.343008
$P(1, 27)$	0.	0.	0.	-0.000009	-0.001354	0.036768	0.270429	0.531541	0.357021
$P(1, 29)$	0.	0.	0.	-0.000001	-0.000215	0.009705	0.118777	0.447103	0.214103
$P(1, 31)$	0.	0.	0.	0.	-0.000029	0.002143	0.041885	0.265834	0.072427
$P(1, 33)$	0.	0.	0.	0.	-0.000003	0.000404	0.012257	0.122961	0.015270
$P(1, 35)$	0.	0.	0.	0.	-0.000000	0.000066	0.003045	0.046442	0.002166
$P(1, 37)$	0.	0.	0.	0.	0.	0.000009	0.000653	0.014757	0.000218
$P(1, 39)$	0.	0.	0.	0.	0.	0.000001	0.000122	0.004027	0.000016
$P(1, 41)$	0.	0.	0.	0.	0.	0.000000	0.000020	0.000958	0.000001
$P(1, 43)$	0.	0.	0.	0.	0.	0.	0.000003	0.000201	0.000000
$P(1, 45)$	0.	0.	0.	0.	0.	0.	0.000000	0.000038	0.000000
$P(1, 47)$	0.	0.	0.	0.	0.	0.	0.	0.000006	0.000000
$P(1, 49)$	0.	0.	0.	0.	0.	0.	0.	0.000001	0.000000
$P(1, 51)$	0.	0.	0.	0.	0.	0.	0.	0.000000	0.000000
h (km)	-12.2703	-1.7581	-0.6443	-0.3297	-0.1996	-0.1337	-0.0957	-0.0719	

TABLE 3.5

Expansion coefficients relating normalized Hough functions $\Theta(L,M)$ and normalized associated Legendre functions $P(L,N)$ Anti-symmetric gravitational modes (index M positive) with wave number $L = 1$ and period $S1 = 24.0000$ mean solar hours. Also shown are equivalent depths

	$\Theta(1,2)$	$\Theta(1,4)$	$\Theta(1,6)$	$\Theta(1,8)$	$\Theta(1,10)$	$\Theta(1,12)$	$\Theta(1,14)$	$\Theta(1,16)$	Sum of squares
$P(1, 2)$	0.000737	0.000303	−0.000174	0.000117	0.000085	−0.000065	0.000052	0.000043	0.000001
$P(1, 4)$	0.287901	0.118432	−0.068122	0.045529	0.033153	−0.025520	0.020424	0.016825	0.106077
$P(1, 6)$	−0.577410	−0.211881	0.118510	−0.078352	−0.056751	0.043555	−0.034795	−0.028628	0.405628
$P(1, 8)$	0.601351	0.108899	−0.046182	0.026793	0.018090	−0.013317	0.010360	0.008372	0.377015
$P(1, 10)$	−0.414543	0.171741	−0.126332	0.090403	0.067789	−0.052994	0.042804	0.035470	0.235967
$P(1, 12)$	0.207905	−0.444406	0.248539	−0.160917	−0.114910	0.087389	−0.069392	−0.056856	0.357276
$P(1, 14)$	−0.079908	0.553020	−0.190225	0.092252	0.054832	−0.036913	0.026944	0.020789	0.362440
$P(1, 16)$	0.024353	−0.487081	−0.036873	0.085619	0.080481	−0.068991	0.058461	0.049842	0.263670
$P(1, 18)$	−0.006033	0.336144	0.302389	−0.239040	−0.176232	0.134469	−0.106472	−0.086900	0.329635
$P(1, 20)$	0.001239	−0.190455	−0.473269	0.244772	0.137755	−0.086672	0.059470	0.043532	0.352092
$P(1, 22)$	−0.000214	0.091042	0.499816	−0.084263	0.023403	−0.050489	0.054909	0.052534	0.274077
$P(1, 24)$	0.000032	−0.037394	−0.416387	−0.157342	−0.198145	0.170912	−0.140996	−0.116846	0.301539
$P(1, 26)$	−0.000004	0.013375	0.290137	0.364661	0.265969	−0.174093	0.118042	0.084105	0.339392
$P(1, 28)$	0.000000	−0.004210	−0.174316	−0.465027	−0.178613	0.045705	0.007926	0.029264	0.281565
$P(1, 30)$	−0.000000	0.001176	0.092012	0.452631	−0.020194	0.134247	−0.146592	−0.134567	0.271369
$P(1, 32)$	0.000000	−0.000294	−0.043224	−0.367164	0.237404	−0.252664	0.194127	0.143562	0.315173
$P(1, 34)$	−0.000000	0.000066	0.018244	0.258183	−0.390932	0.238383	−0.110313	−0.040447	0.290451
$P(1, 36)$	0.	−0.000013	−0.006970	−0.160909	0.445628	−0.097434	−0.056745	−0.105754	0.248422
$P(1, 38)$	0.	−0.000002	0.002425	0.090162	−0.413204	−0.104527	0.208488	0.193602	0.270747
$P(1, 40)$	0.	−0.000000	−0.000772	−0.045879	0.330159	0.287880	−0.260199	−0.161095	0.287641
$P(1, 42)$	0.	0.000000	0.000226	0.021361	−0.233984	−0.398678	0.185722	0.023872	0.249211
$P(1, 44)$	0.	0.	−0.000061	−0.009154	0.149637	0.423328	−0.020415	0.140875	0.221944
$P(1, 46)$	0.	0.	0.000015	0.003628	−0.087351	−0.380152	−0.168436	−0.245144	0.240625
$P(1, 48)$	0.	0.	−0.000004	−0.001335	0.046930	0.300892	0.318640	0.238180	0.251001
$P(1, 50)$	0.	0.	0.000001	0.000458	−0.023351	−0.214661	−0.396634	−0.125141	0.219603
$P(1, 52)$	0.	0.	−0.000000	−0.000147	0.010814	0.139973	0.401037	−0.046266	0.182681
$P(1, 54)$	0.	0.	0.	0.000044	−0.004680	−0.084223	−0.352066	0.215560	0.177532
$P(1, 56)$	0.	0.	0.	−0.000012	0.001899	0.047093	0.276908	−0.336400	0.192064
$P(1, 58)$	0.	0.	0.	0.000003	−0.000724	−0.024600	−0.198671	0.389326	0.191651
$P(1, 60)$	0.	0.	0.	−0.000001	0.000260	0.012057	0.131546	−0.379800	0.161698

Table 3.5 (continued)

	$\Theta(1,2)$	$\Theta(1,4)$	$\Theta(1,6)$	$\Theta(1,8)$	$\Theta(1,10)$	$\Theta(1,12)$	$\Theta(1,14)$	$\Theta(1,16)$	Sum of squares
$P(1, 62)$	0.	0.	0.	0.000000	−0.000088	−0.005563	−0.081040	0.327857	0.114089
$P(1, 64)$	0.	0.	0.	−0.000000	0.000028	0.002424	0.046734	−0.256730	0.068100
$P(1, 66)$	0.	0.	0.	0.	−0.000009	−0.001000	−0.025346	0.185097	0.034904
$P(1, 68)$	0.	0.	0.	0.	0.000003	0.000391	0.012976	−0.124096	0.015568
$P(1, 70)$	0.	0.	0.	0.	−0.000001	−0.000145	−0.006291	0.077915	0.006110
$P(1, 72)$	0.	0.	0.	0.	0.000000	0.000052	0.002895	−0.046058	0.002130
$P(1, 74)$	0.	0.	0.	0.	−0.000000	−0.000017	−0.001268	0.025740	0.000664
$P(1, 76)$	0.	0.	0.	0.	0.	0.000006	0.000529	−0.013646	0.000186
$P(1, 78)$	0.	0.	0.	0.	0.	−0.000002	−0.000211	0.006881	0.000047
$P(1, 80)$	0.	0.	0.	0.	0.	0.000001	0.000080	−0.003309	0.000011
$P(1, 82)$	0.	0.	0.	0.	0.	−0.000000	−0.000029	0.001520	0.000002
$P(1, 84)$	0.	0.	0.	0.	0.	0.000000	0.000010	−0.000668	0.000000
$P(1, 86)$	0.	0.	0.	0.	0.	0.	−0.000003	0.000282	0.000000
$P(1, 88)$	0.	0.	0.	0.	0.	0.	0.000001	−0.000114	0.000000
$P(1, 90)$	0.	0.	0.	0.	0.	0.	−0.000000	0.000044	0.000000
$P(1, 92)$	0.	0.	0.	0.	0.	0.	0.000000	−0.000017	0.000000
$P(1, 94)$	0.	0.	0.	0.	0.	0.	−0.000000	0.000006	0.000000
$P(1, 96)$	0.	0.	0.	0.	0.	0.	0.	−0.000002	0.000000
$P(1, 98)$	0.	0.	0.	0.	0.	0.	0.	0.000001	0.000000
$P(1, 100)$	0.	0.	0.	0.	0.	0.	0.	−0.000000	0.000000
$P(1, 102)$	0.	0.	0.	0.	0.	0.	0.	0.000000	0.000000
$P(1, 104)$	0.	0.	0.	0.	0.	0.	0.	−0.000000	0.000000
h (km)	0.2384	0.0724	0.0346	0.0202	0.0132	0.0093	0.0069	0.0054	

TABLE 3.6

Expansion coefficients relating normalized Hough functions $\Theta(L,M)$ and normalized associated Legendre functions $P(L,N)$. Anti-symmetric rotational modes (index M negative) with wave number $L = 1$ and period $S1 = 24.0000$ mean solar hours. Also shown are equivalent depths

	$\Theta(1,-1)$	$\Theta(1,-3)$	$\Theta(1,-5)$	$\Theta(1,-7)$	$\Theta(1,-9)$	$\Theta(1,-11)$	$\Theta(1,-13)$	$\Theta(1,-15)$	Sum of Squares
$P(1, 2)$	0.999997	0.002094	-0.000896	-0.000529	0.000360	-0.000266	-0.000206	0.000166	0.999999
$P(1, 4)$	-0.002556	0.819988	-0.350227	-0.206831	0.140724	-0.103751	-0.080587	0.064936	0.879103
$P(1, 6)$	0.000002	0.551233	0.298376	0.261514	-0.201818	0.157718	0.126493	-0.103945	0.553686
$P(1, 8)$	0.	0.152274	0.728831	0.242549	-0.085917	0.029445	0.006687	0.003106	0.621515
$P(1, 10)$	0.	0.023753	0.476220	-0.413472	0.368516	-0.277678	-0.211086	0.165423	0.683139
$P(1, 12)$	0.	0.002409	0.169513	-0.660846	0.112416	0.075972	0.128425	-0.136329	0.518945
$P(1, 14)$	0.	0.000172	0.039433	-0.434126	-0.454815	0.377257	0.240640	-0.147719	0.622216
$P(1, 16)$	0.	0.000009	0.006559	-0.176554	-0.610048	0.009868	-0.198311	0.235815	0.498406
$P(1, 18)$	0.	0.	0.000824	-0.050650	-0.406187	-0.474516	-0.344389	0.155109	0.535383
$P(1, 20)$	0.	0.	0.000081	-0.010968	-0.180510	-0.570862	0.065774	-0.275223	0.438662
$P(1, 22)$	0.	0.	0.000006	-0.001871	-0.059366	-0.385800	0.477702	-0.297400	0.469015
$P(1, 24)$	0.	0.	0.000000	-0.000259	-0.015269	-0.183018	0.539646	0.121192	0.339635
$P(1, 26)$	0.	0.	0.	-0.000030	-0.003182	-0.066432	0.370006	0.474765	0.366729
$P(1, 28)$	0.	0.	0.	-0.000003	-0.000551	-0.019342	0.184705	0.514096	0.298785
$P(1, 30)$	0.	0.	0.	-0.000000	-0.000081	-0.004658	0.072323	0.357265	0.132891
$P(1, 32)$	0.	0.	0.	0.	-0.000010	-0.000948	0.023155	0.185882	0.035089
$P(1, 34)$	0.	0.	0.	0.	-0.000001	-0.000166	0.006228	0.077340	0.006020
$P(1, 36)$	0.	0.	0.	0.	0.	-0.000025	0.001435	0.026712	0.000716
$P(1, 38)$	0.	0.	0.	0.	0.	-0.000003	0.000287	0.007849	0.000062
$P(1, 40)$	0.	0.	0.	0.	0.	-0.000000	0.000051	0.001997	0.000004
$P(1, 42)$	0.	0.	0.	0.	0.	-0.000000	0.000008	0.000446	0.000000
$P(1, 44)$	0.	0.	0.	0.	0.	0.	0.000001	0.000088	0.000000
$P(1, 46)$	0.	0.	0.	0.	0.	0.	0.000000	0.000016	0.000000
$P(1, 48)$	0.	0.	0.	0.	0.	0.	0.	0.000002	0.000000
$P(1, 50)$	0.	0.	0.	0.	0.	0.	0.	0.000000	0.000000
h (km)	+803.356	-1.8144	-0.6460	-0.3298	-0.1996	-0.1337	-0.0457	-0.07193	

TABLE 3.7

Expansion coefficients relating normalized Hough functions $\Theta(L,M)$ and normalized associated Legendre functions $P(L,N)$. Symmetric gravitational modes (index M positive) with wave number $L = 2$ and period $M2 = 12.4206$ mean solar hours. Also shown are equivalent depths

	$\Theta(2,2)$	$\Theta(2,4)$	$\Theta(2,6)$	$\Theta(2,8)$	$\Theta(2,10)$	$\Theta(2,12)$	$\Theta(2,14)$	$\Theta(2,16)$	Sum of squares
$P(2, 2)$	0.962708	0.228431	0.105893	−0.062925	−0.042627	−0.031293	−0.024243	0.019515	0.997926
$P(2, 4)$	−0.268877	0.750658	0.421970	−0.266311	−0.185291	−0.137970	−.07795	0.087244	0.957362
$P(2, 6)$	0.029910	−0.594197	0.342065	−0.364523	−0.301338	−0.243608	−0.199354	0.166071	0.821321
$P(2, 8)$	−0.001739	0.174547	−0.723269	0.092897	−0.119028	−0.175832	−0.181789	0.171503	0.669762
$P(2, 10)$	0.000062	−0.028058	0.395860	0.597957	0.369596	0.159536	0.040166	0.022334	0.679210
$P(2, 12)$	−0.000001	0.002888	−0.115800	−0.587225	0.270947	0.386515	0.311063	−0.219448	0.725975
$P(2, 14)$	0.	−0.000208	0.021874	0.272972	−0.633221	−0.104646	0.179044	−0.260771	0.586971
$P(2, 16)$	0.	0.000011	−0.002931	−0.080173	0.454998	−0.478966	−0.352202	0.104247	0.577781
$P(2, 18)$	0.	−0.000000	0.000295	0.016672	−0.193480	0.576213	−0.171575	0.366883	0.533775
$P(2, 20)$	0.	0.	−0.000023	−0.002616	0.057199	−0.347925	0.554611	−0.155773	0.456189
$P(2, 22)$	0.	0.	0.000001	0.000323	−0.012698	0.140031	−0.491331	−0.365341	0.394650
$P(2, 24)$	0.	0.	0.	−0.000032	0.002218	−0.041690	0.265467	0.550452	0.375213
$P(2, 26)$	0.	0.	0.	0.000003	−0.000315	0.009711	−0.102983	−0.405622	0.175229
$P(2, 28)$	0.	0.	0.	−0.000000	0.000037	−0.001835	0.030868	0.202904	0.042126
$P(2, 30)$	0.	0.	0.	0.	−0.000004	0.000288	−0.007462	−0.076688	0.005937
$P(2, 32)$	0.	0.	0.	0.	0.000000	−0.000038	0.001497	0.023131	0.000537
$P(2, 34)$	0.	0.	0.	0.	0.	0.000004	−0.000255	−0.005761	0.000033
$P(2, 36)$	0.	0.	0.	0.	0.	−0.000000	0.000037	0.001213	0.000001
$P(2, 38)$	0.	0.	0.	0.	0.	0.	−0.000005	−0.000220	0.000000
$P(2, 40)$	0.	0.	0.	0.	0.	0.	0.000001	0.000035	0.000000
$P(2, 42)$	0.	0.	0.	0.	0.	0.	0.	−0.000005	0.000000
$P(2, 44)$	0.	0.	0.	0.	0.	0.	0.	0.000001	0.000000
$P(2, 46)$	0.	0.	0.	0.	0.	0.	0.	−0.000000	0.000000
h (km)	7.0701	1.8486	0.8254	0.4632	0.2953	0.2042	0.1495	0.1141	

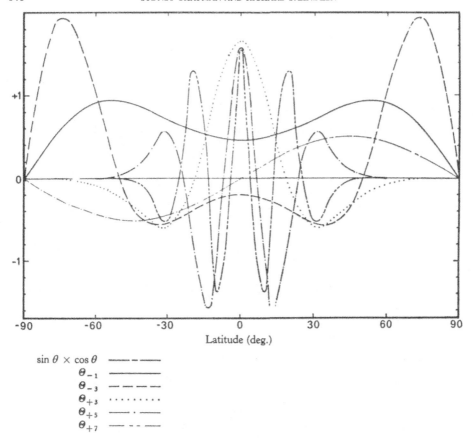

Fig. 3.10.   Symmetric Hough functions for the migrating solar diurnal thermal tide. Also shown is $\sin\theta\cos\theta$, the most important odd mode. After Lindzen (1967a). (Correction to subscripts: -1, -3, $+3$, $+5$, $+7$ should be -2, -4, $+1$, $+3$, $+5$.)

in turn, implies that the modes associated with positive h's will propagate vertically with short wavelengths. For an isothermal atmosphere with $T_0 = 260°$, wavelengths of 28, 11, and 7 km are associated with $h_1^{\omega,\,1}$, $h_3^{\omega,\,1}$ and $h_5^{\omega,\,1}$ respectively (Lindzen, 1967a). These estimates are slightly modified when more realistic basic temperatures are considered (Lindzen, 1968a).

Continuing as in Section 3.5A we expand $g_{O_3}^{\omega,\,1}$ and $g_{H_2O}^{\omega,\,1}$ (in degrees Kelvin) in terms of the relevant Hough Functions:

$$g_{O_3}^{\omega,\,1} = 1.6308\,\mathrm{K}\,\Theta_{-2}^{\omega,\,1} - 0.5128\,\mathrm{K}\,\Theta_{-4}^{\omega,\,1} + \cdots$$
$$+ 0.5447\,\mathrm{K}\,\Theta_1^{\omega,\,1} - 0.1411\,\mathrm{K}\,\Theta_3^{\omega,\,1} + 0.0723\,\mathrm{K}\,\Theta_5^{\omega,\,1} - \cdots \qquad (168)$$
$$g_{H_2O}^{\omega,\,1} = 0.157\,\mathrm{K}\,\Theta_{-2}^{\omega,\,1} - 0.055\,\mathrm{K}\,\Theta_{-4}^{\omega,\,1} \cdots$$
$$+ 0.062\,\mathrm{K}\,\Theta_1^{\omega,\,1} - 0.016\,\mathrm{K}\,\Theta_3^{\omega,\,1} + 0.008\,\mathrm{K}\,\Theta_5^{\omega,\,1} \cdots. \qquad (169)$$

We can immediately note the absence in (168) and (169) of the strong dominance of one mode *vis a vis* the others – in distinct contrast to (147) and (150). The reason

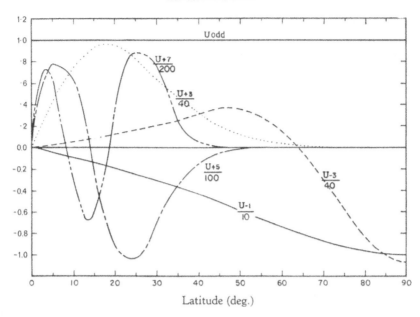

Fig. 3.11. The expansion functions for the latitude dependence of the solar diurnal component of
u, the northerly velocity. The functions have been divided by the amounts shown. After Lindzen
(1967a). (Correction to subscripts: -1, -3, $+3$, $+5$, $+7$ should be -2, -4, $+1$, $+3$, $+5$.)

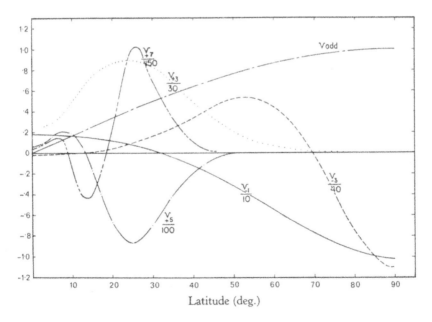

Fig. 3.12. The expansion functions for the latitude dependence of the solar diurnal component of *v*,
the westerly velocity. The functions have been divided by the amounts shown. After Lindzen (1967a).
(Correction to subscripts: -1, -3, $+3$, $+5$, $+7$ should be -2, -4, $+1$, $+3$, $+5$.)

for this is seen in Figures 3.10 and 3.2; no single diurnal Hough mode bears much resemblance to the excitation distributions, though the $\Theta_{-2}^{\omega, 1}$ mode comes closer than any other. However, since this mode has a negative equivalent depth, it is not – in contrast to the $\Theta_{2}^{\omega, 2}$ mode – associated with a preferential atmospheric response. This is seen more clearly in the theoretically calculated diurnal surface pressure response to the excitations shown in Figure 3.2 (for an isothermal basic state). The results are taken from Lindzen (1967a). The oscillations excited by O_3 are considered separately from those excited by H_2O.

$$\delta p_{H_2O}(0) = \{137\Theta_{-2}^{\omega, 1} - 68.2\Theta_{-4}^{\omega, 1} + 117\, e^{56°i}\Theta_1^{\omega, 1}$$
$$- 13.0e^{75.3°i}\, \Theta_3^{\omega, 1} + 4.11e^{80.5°i}\, \Theta_5^{\omega, 1} + \cdots\} e^{i(\omega t + \phi)}\, \mu b \qquad (170)$$

and

$$\delta p_{O_3}(0) = \{44.1\, \Theta_{-2}^{\omega, 1} - 3.4\, \Theta_{-4}^{\omega, 1} + 94.1e^{12.75°i}\, \Theta_1^{\omega, 1}$$
$$- 3.75e^{16.1°i}\, \Theta_3^{\omega, 1} + 0.754e^{-6.57°i}\, \Theta_5^{\omega, 1} + \cdots\} e^{i(\omega t + \phi)}\, \mu b. \qquad (171)$$

We see from (170) and (171) that at least 3 modes, $\Theta_{-2}^{\omega, 1}$, $\Theta_1^{\omega, 1}$, and $\Theta_{-4}^{\omega, 1}$, and not just one, are of significance in the surface pressure oscillation. We also see that although the $\Theta_{-2}^{\omega, 1}$ mode receives most of the excitation, the $\Theta_1^{\omega, 1}$ mode plays as great a role in $\delta p_{H_2O}^{(0)}$ and a greater role in $\delta p_{O_3}^{(0)}$. This is because the $\Theta_{-2}^{\omega, 1}$ mode is associated with a negative equivalent depth, and energy in this mode cannot propagate away from a source – it can only leak away. This trapping becomes increasingly great for the modes $\Theta_{-4}^{\omega, 1}$, $\Theta_{-6}^{\omega, 1}$. Thus we are not surprised that the contributions to the modes with negative h's from water vapor (near the ground) are larger than the contributions from ozone (far above the ground). However, the contributions from water vapor absorption to the modes with positive h's are also larger. This is due to the short vertical wavelengths associated with these modes. The ozone excitation is distributed over a very considerable depth of the atmosphere (ca. 40 km). Thus, waves excited at one level can destructively interfere with waves excited at another level (see Butler and Small (1963), and Lindzen (1966b), for a more detailed discussion of this process). For the $\Theta_1^{\omega, 1}$ mode (wavelength ~ 28 km) the region of water vapor excitation is not sufficiently thick (ca. 18 km) for this process to be of great importance. This, however, is no longer true for the $\Theta_3^{\omega, 1}$ and subsequent modes.

The above discussion explains why the migrating solar diurnal surface pressure oscillation is relatively weak – namely inefficiency of response due to trapping and interference. It also explains why it is more irregular than the semidiurnal surface pressure oscillation – namely the oscillation consists in several modes, each of which is sensitive to relatively local variations in excitation, temperature, etc.

It may be asked at this point whether the above implies that the migrating solar diurnal thermotidal fields are of secondary importance throughout the atmosphere. The answer is no. The reasons for this are (1) the solar diurnal excitation is very large; even the excitation received by the modes with positive h's alone is comparable with the solar semidiurnal excitation; (2) trapping and interference do not preclude an effective response at the levels of excitation; (3) those modes which propagate vertically are not trapped below the mesopause. The last item suggests that diurnal thermotidal fields will dominate semidiurnal fields in the upper atmosphere equatorwards of about ±40°. This is seen more clearly in Figures 3.13 and 3.14, where theoretical predictions for the amplitude and phase of the solar diurnal northerly velocity component at various latitudes are shown. The figures are taken from Lindzen (1967a), who used the five Hough modes shown in Figure 3.10, an isothermal basic

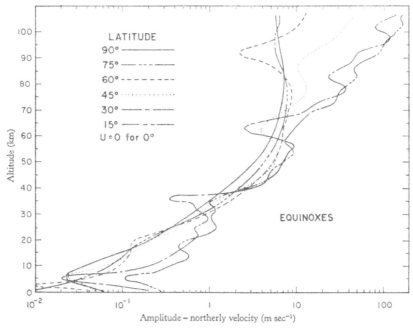

Fig. 3.13. Altitude distribution of the amplitude of the solar diurnal component of u at 15° intervals of latitude; isothermal basic state assumed. After Lindzen (1967a).

state where $T_0 = 260°$K, and the excitations shown in Figure 3.2. Figures 3.13 and 3.14 suggest a very rich latitude and altitude structure. A more explicit display of the calculated latitude structure is shown in Figures 3.15 and 3.16 where the amplitude and phase of the diurnal temperature and velocity fields at 85 km as a function of latitude are shown. The vertical structure is best seen in Figure 3.17 showing the migrating solar diurnal contribution to the altitude distribution of northerly velocity at 1200 hrs and 1800 hrs, and at ±30° and ±50°. (Figures 3.15–17 are also taken from Lindzen (1967a).) We see in Figure 3.17 that at ±30° the vertical structure and amplitude are much greater than at ±50° – consistent with the dominance of trapped modes at higher latitudes. The structure and amplitude of the diurnal contributions above 80 km at ±30° are both very similar to single time observations of the total wind at 30° (Liller and Whipple, 1954) – thus suggesting that the migrating solar diurnal thermotidal winds may be the major component of the total wind above 80 km at latitudes equatorwards of about 30°.

Figures 3.13–17 are based on calculations for an isothermal basic state. The results

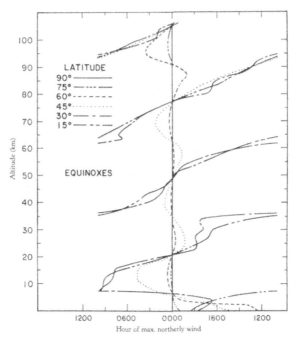

Fig. 3.14. Altitude distribution of the phase of the solar diurnal component of u at 15° intervals of latitude; isothermal basic state assumed. After Lindzen (1967a).

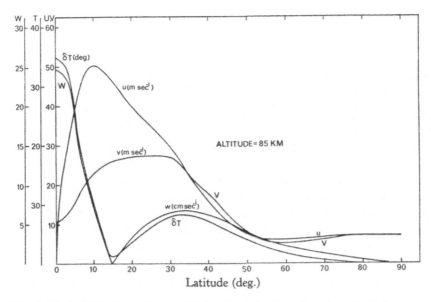

Fig. 3.15. Latitude distribution of the amplitude of the solar diurnal components of the u, v, w and T fields at 85 km; isothermal basic state assumed. After Lindzen (1967a).

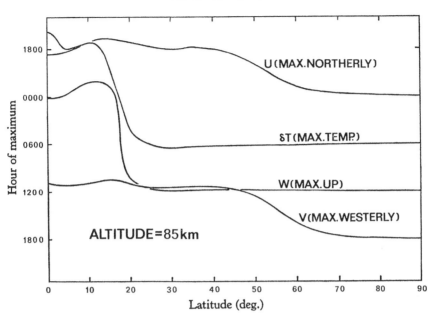

Fig. 3.16. Latitude distribution of the phase of the solar diurnal components of the u, v, w and T fields at 85 km; isothermal basic state assumed. After Lindzen (1967a).

of these calculations are in reasonable agreement with various observations (Lindzen, 1967a; see also Section 2S.7B). This, however, is not, by itself, a justification for the use of an isothermal atmosphere. A weak justification consists in noting that for negative h's or small positive h's, λ^2 does not change sign anywhere in the atmosphere. Thus the attenuations and reflections associated with the transition between regions of propagation and evanescence (viz. Section 3.5.A) are not to be expected for diurnal oscillations. On the other hand there are normal temperature variations on the scale of the wavelengths of the main diurnal modes, and refraction effects would not be surprising. Lindzen (1968a) has investigated the effect of various basic temperature profiles on the solar diurnal thermal tide and has found very little effect on the surface pressure oscillation. However, more important effects were predicted for upper air fields. In Figure 3.18 we see several temperature profiles differing only in the lower atmosphere. In Figure 3.19 we see the predicted diurnal component of the northerly velocity as a function of altitude at 1200 hrs (local time) over 25° latitude for each of the temperature profiles in Figure 3.18. Small changes in T_0 appear to have an important influence on the relative importance of the various propagating diurnal modes.

3.5.C. THE LUNAR SEMIDIURNAL TIDE

For the semidiurnal lunar tide (M_2) $s = 2$ and $f = 12.0/12.4206$. The solutions of (69) and (70) for these values have been known for some time; again, excellent tabulations are to be found in Flattery (1967). Following his notation, $n = 2, 4, \ldots$ correspond to

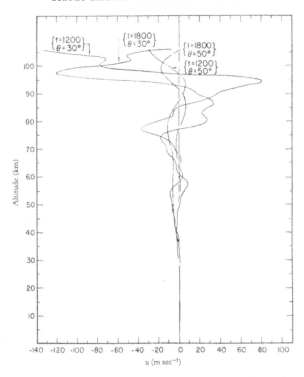

Fig. 3.17.   Altitude distribution of solar diurnal contribution to u at different times (1200 and 1800 hours local time) and at different latitudes (30° and 50°). After Lindzen (1967a).

symmetric modes, and, as we saw in Section 3.4.A, only symmetric modes are excited; $\{h_n\}$ and $\{\bar{C}_{n,m}\}$ for these modes are given in Table 3.7. Because of the substantial similarity between the Θ, U and V functions for the lunar and solar semidiurnal oscillations we have not included illustrations of the lunar examples.

As in the preceding cases, our first step is to expand our excitation in terms of the appropriate Hough functions, using (129)

$$\Omega = (-23662.\,\Theta_2 - 5615.\,\Theta_4 - 2603.\,\Theta_6 \cdots)\cos\left(2(\sigma_2^L t + \varphi)\right)\mathrm{cm}^2/\mathrm{sec}^2.$$
$$(172)$$

Just as with the solar semidiurnal thermal tide, most of the excitation goes into the $n=2$ mode. For gravitational tides $J \equiv 0$, and Equation (30) is homogeneous. The excitation enters only through the lower boundary condition (47). Thus for gravitational tides the drive is, in effect, concentrated at a single level – the ground. Restricting himself to the $n=2$ mode, Sawada (1954, 1956) has computed the atmosphere's response to (172) for various distributions of T_0. In Figure 3.20 we see various distributions of T_0 differing only in their mesopeak temperatures. In Figure 3.21 we see the phase and amplitude of the surface pressure oscillation at the equator for the various distributions of T_0. In marked contrast to the results for the solar semidiurnal thermal tide, the lunar semidiurnal tide in the surface pressure seems extremely

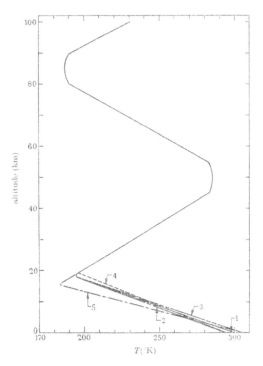

Fig. 3.18. Different basic temperature profiles used in examining the diurnal thermal tide. After
Lindzen (1968a).

dependent on the precise distribution of T_0. In view of the similarities in the
equivalent depths and Hough functions for the two oscillations this difference
is quite surprising. The difference can only arise from the difference between the
excitations: the excitation for the lunar semidiurnal tide is a coherent drive at a
single level, while the excitation for the solar semidiurnal thermal tide is distributed
throughout a great depth of the atmosphere. Apparently, small changes in the
distribution of T_0 change the effective height of levels where semidiurnal tides are
partially reflected and the degrees of reflection. For the coherent gravitational excitation, the repeated partial reflections seem to produce significant constructive or
destructive interference depending on the height of the reflecting levels. For the
distributed thermal excitation, constructive interference for waves excited at one level
appears to be balanced by destructive interference for waves excited at another level –
the net variation being small.

Sawada (1956) attempted to use the sensitivity of the lunar semidiurnal surface
pressure oscillation to $T_0(z)$ in order to determine the annual mean distribution of
T_0. Presumably it would be that $T_0(z)$ for which the pressure oscillation was closest to
the observed one (i.e., $|\delta p| \approx 70$ μb, phase $(\delta p) \approx 72°$). Thus Sawada concluded that
T_0 should have a maximum value of $262.33°$K at 50 km. Such a conclusion seems
unwarranted on several grounds. For example, consider an atmosphere for which

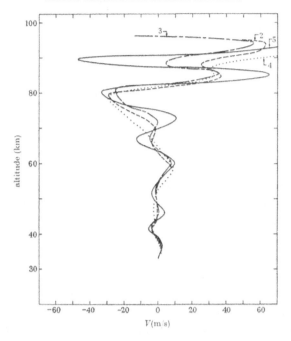

Fig. 3.19.   Solar diurnal component of v at 1200 LST for latitude 25°. The different distributions
correspond to different basic temperature profiles. After Lindzen (1968a).

T_0 =const. Restricting ourselves to the $n=2$ mode, we have as the solution of (48),

$$y_2 = A e^{i\lambda z/H},$$ (173)

where

$$\lambda = \left(\frac{\kappa H}{h_2} - \frac{1}{4}\right)^{1/2}.$$ (174)

From (47)

$$A = \left(\frac{i\sigma \Omega_2}{\gamma g h_2}\right) \bigg/ \left(i\lambda + \left(\frac{H}{h_2} - \frac{1}{2}\right)\right),$$ (175)

Finally, from (35) and (172)

$$\delta p_{\text{surface}} \approx \left[\frac{34.17\mu b \exp i\left(2(\sigma_2^{\text{L}} t + \phi) + 90°\right)}{\left(\left(\frac{H}{h} - \frac{1}{2}\right) + i \sqrt{\frac{\kappa H}{h} - \frac{1}{4}}\right)}\right].$$ (176)

For $T_0 = 230°\text{K}$ (i.e., $H = 6.76$ km), (176) represents the observations quite well. Yet
we could hardly conclude that the atmosphere is isothermal at 230°K. The above
result is of some interest, since waves will not be reflected at any level in an isothermal
atmosphere. It may, therefore, be suggested that the multiple reflections producing
Sawada's result are, on the average, unimportant. A possible reason for this will be
given in Section 6, where we will show that the effects of multiple reflections are

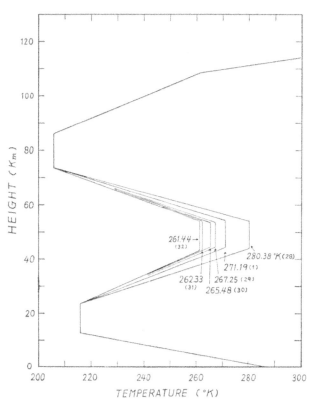

Fig. 3.20. Various temperature profiles used in calculating the lunar semidiurnal surface pressure oscillation. The maximum temperature of the ozonosphere and a profile number are shown for each of the profiles. After Sawada (1956).

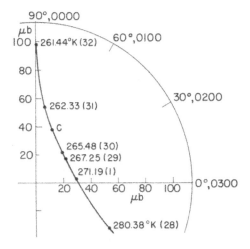

Fig. 3.21. A harmonic dial for the lunar semidiurnal surface pressure oscillation. Amplitude and phase are shown as functions of the basic temperature profile. After Sawada (1956).

greatly reduced when dissipation is included. Despite the mitigating role of dissipation, dependence on $T_0(z)$ is still sufficient to account for the observed variability in the lunar semidiurnal tide.

As Sawada (1956) noted, atmospheric fields above 25 km are considerably less sensitive than surface pressure to small changes in $T_0(z)$. In Figures 3.22 and 3.23 we show the results of some of our calculations for the distribution with height of the amplitude and phase of the lunar semidiurnal component of the westerly wind at the equator as computed for the ARDC temperature profile. For comparison we also

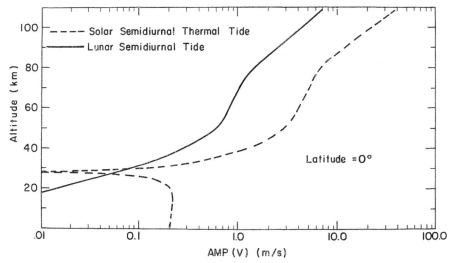

Fig. 3.22.   Amplitudes of lunar and solar semidiurnal contributions to the westerly velocity over the equator; ARDC temperature profile assumed.

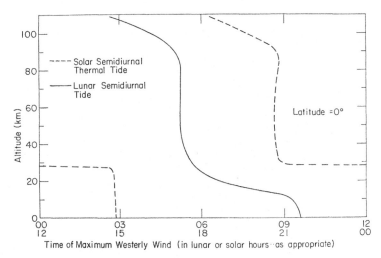

Fig. 3.23.   Phases of the lunar and solar semidiurnal contributions to the westerly velocity v over the equator; ARDC temperature profile assumed.

show the contributions from the solar semidiurnal thermal tide – assuming the same temperature profile. Both are seen to vary with height in a similar manner.

3.5.D. OTHER COMPONENTS

We have discussed the three most important tides and thermal tides. The literature contains theoretical discussions of lesser components. In particular, both Butler and Small (1963) and Nunn (1967) discuss the solar terdiurnal thermal tide. Lindzen (1967c) discusses the lunar diurnal tide. A discussion of the standing solar semidiurnal thermal tide may be found in Stolov (1954) and Kertz (1956a).

3.6. Shortcomings of Present Calculations

In this section we return to a discussion of the approximations g to j, cited in Section 3.2: namely, the effects on tides and thermal tides of surface topography, dissipation, nonlinearity, and mean flows. Of necessity, this section will be not so much a review as a discussion of conjectures and of work which has yet to be done. The amount of work already done on these approximations is small, and, for the most part, inconclusive. However, in a few instances (such as the effect of dissipation on the lunar semidiurnal tide) concrete results have already been obtained.

3.6.A. SURFACE TOPOGRAPHY

We have, in our theoretical development, assumed the earth to be a perfectly smooth sphere. This led to the lower boundary condition given by Equation (46) – namely

$$w = 0 \quad \text{at} \quad z = 0.$$

Our analysis could be extended to take account of small, smooth variations in surface elevation. Let $\xi(\Theta, \phi)$ be the elevation of the surface. If this can be considered a perturbation about a smooth surface, then (46) can be replaced by

$$w = \mathbf{u}_{\text{hor}} \cdot \nabla \xi \quad \text{at} \quad z = 0, \tag{177}$$

(see Charney and Eliassen (1949); Eliassen and Palm (1961), for details). Let us rewrite (177) as follows:

$$w = \varepsilon(a(\theta, \phi)u + b(\theta, \phi)v), \tag{178}$$

where $\varepsilon \ll 1$ and $a \sim O(1)$, $b \sim O(1)$. Let our excitation have a time and longitude variation of the form

$$e^{i(\sigma t + s\phi)} \tag{179}$$

and let

$$a = \sum_{\substack{-\infty \\ n \neq 0}}^{\infty} a_n(\theta) e^{in\phi} \tag{180}$$

$$b = \sum_{\substack{-\infty \\ n \neq 0}}^{\infty} b_n(\theta) e^{in\phi}. \tag{181}$$

The quantity $\sigma a/s$ is the speed at which the excitation travels around the earth at the equator.

Let us now take our final solution to be an expansion in ε, i.e.,

$$
\begin{pmatrix} w \\ u \\ v \\ \vdots \end{pmatrix} = \begin{pmatrix} w_0 \\ u_0 \\ v_0 \\ \vdots \end{pmatrix} + \varepsilon \begin{pmatrix} w_1 \\ u_1 \\ v_1 \\ \vdots \end{pmatrix} + \varepsilon^2 \begin{pmatrix} w_2 \\ u_2 \\ v_2 \\ \vdots \end{pmatrix} + \cdots .
\tag{182}
$$

The zeroth-order solution will merely be our original solution for which $W_0 = 0$ at $Z = 0$. This solution will travel around the earth at the same speed as the excitation. To first order in ε we have

$$
\begin{aligned}
W_1 = u_0(0) e^{i\sigma t} \sum_{\substack{n=-\infty \\ n \neq 0}}^{\infty} a_n(\theta) e^{i(n+s)l} \\
+ v_0(0) e^{i\sigma t} \sum_{\substack{n=-\infty \\ n \neq 0}}^{\infty} b_n(\theta) e^{i(n+s)l} \quad \text{at} \quad z = 0 .
\end{aligned}
\tag{183}
$$

Thus, to first order in ε we generate a set of tidal oscillations which do not trave with the excitation. These will create longitudinal variations in the tides. However to first order in ε we do not affect the zeroth order solution which does travel with the excitation. To second order in ε, both the oscillations which follow and do not follow the excitation are affected. Thus to second order in ε we will also influence the resonance properties of the atmosphere (viz. Section 3.5.A). This last feature has been investigated by Kertz (1951).

So far we have considered only the kinematical effects of surface topography. However, there may also be longitudinal variations in insolation absorbing gases due to the distribution of land and sea. This will be particularly important for water vapor (and, according to Hinzpeter (private communication), for insolation-absorbing aerosols), and, hence, for the diurnal solar thermal tide (viz. Section 3.5.B). In a more obvious sense it is also true for the most important absorber of sunlight – land itself. Roughly speaking, the effect of longitudinal variations in the distribution of insolation absorption will be to replace (143) which can be written

$$
J = \hat{J}(z, \theta) e^{in(\sigma t + \phi)} ,
$$

where $\sigma = 2\pi/1$ solar day, by

$$
J = \hat{J}(z, \theta, \phi) e^{in(\sigma t + \phi)} ,
\tag{184}
$$

where the ϕ-dependence of J reflects the longitude variation in absorbing gases. Let

$$
\hat{J} = J_0(z, \theta) + J_1(z, \theta) \cos(\phi + P_1) + J_2(z, \theta) \cos(2\phi + P_2) + \cdots,
\tag{185}
$$

where $P_1, P_2, \ldots$ are constants; (185) can be rewritten as

$$
\begin{aligned}
\hat{J} = J_0(z, \theta) + \tfrac{1}{2} J_1(z, \theta) \left(e^{i(\phi + P_1)} + e^{-i(\phi + P_1)} \right) \\
+ \tfrac{1}{2} J_2(z, \theta) \left(e^{i(2\phi + P_2)} + e^{-i(2\phi + P_2)} \right) + \cdots .
\end{aligned}
\tag{186}
$$

Substituting (186) into (184) we get

$$J = J_0 \, e^{in(\sigma t + \phi)} + \tfrac{1}{2} J_1 \, e^{iP_1} \, e^{i(n\sigma t + (n+1)\phi)}$$
$$+ \tfrac{1}{2} J_1 \, e^{-iP_1} \, e^{i(n\sigma t + (n-1)\phi)} + \tfrac{1}{2} J_2 \, e^{iP_2} \, e^{i(n\sigma t + (n+2)\phi)}_i$$
$$+ \tfrac{1}{2} J_2 \, e^{-tP_2} \, e^{i(n\sigma t + (n-2)\phi)} + \cdots . \tag{187}$$

Let us now restrict ourselves to the solar diurnal thermal tide, for which $n=1$; (187) becomes

$$J = J_0 \, e^{i(\sigma t + \phi)} + \tfrac{1}{2} J_1 \, e^{iP_1} \, e^{i(\sigma t + 2\phi)}$$
$$+ \tfrac{1}{2} J_1 \, e^{-iP_1} \, e^{i\sigma t} + \tfrac{1}{2} J_2 \, e^{iP_2} \, e^{i(\sigma t + 3\phi)}$$
$$+ \tfrac{1}{2} J_2 \, e^{-iP_2} \, e^{i(\sigma t - \phi)} + \cdots . \tag{188}$$

From (188) we see that J will include not only a component following the sun, but also components which are stationary, which move faster than the sun, which move in a direction opposite to the sun, etc.

The evaluation of the above described effects would ultimately require the evaluation of many additional Hough functions (corresponding to the generation of modes with different zonal wave numbers), and this may, in part, account for the fact that no such evaluation is to be found in the literature. (A preliminary analysis is found in Kertz, 1956a.) However, such evaluations would be quite straightforward. Hopefully, with the aid of electronic computers, they will be forthcoming.

3.6.B. DISSIPATION

The theory of atmospheric tides, as we have presented it so far, has ignored all dissipative-type mechanisms. In the atmosphere, these are turbulence, molecular diffusion of heat and momentum, infrared cooling, and ion drag. Each of these mechanisms is associated with a time scale. For example if we were to include viscosity, Equation (8) would become

$$\frac{\partial u}{\partial t} - 2\omega v \cos\theta = -\frac{1}{a}\frac{\partial}{\partial\theta}\left(\frac{\delta p}{\varrho_0} + \Omega\right) + \frac{v}{\varrho_0} \, \nabla^2 u , \tag{189}$$

where '$\nabla^2 u$' = the proper expression in spherical coordinates for a 'shallow' atmosphere. Let our tide be associated with a vertical wavelength L_Z and a horizontal scale L_H. Then the order of magnitude of $(v/\varrho_0)\,\nabla^2 u$ is given by the larger of $vu/\varrho_0 L_Z^2$ and $vu/\varrho_0 L_H^2$. This is usually $vu/\varrho_0 L_Z^2$. The associated time scale is simply $\varrho_0 L_Z^2/v = \tau_{\mathrm{visc},.}$

Wilkes (1949) claims that a dissipative process will be important when $\tau_{\mathrm{diss.}}$ $< \tau_{\mathrm{tide}}/2\pi$, where τ_{tide} is the period of the tidal oscillation. This is certainly true in a local sense. That is to say, at any given point dissipation will be important if $\tau_{\mathrm{diss.}}$ $< \tau_{\mathrm{tide}}/2\pi$. However, the tides and thermal tides we have dealt with are essentially waves which are excited in the mesosphere and below, and propagate upward. The time it takes these waves to traverse a region where dissipation occurs is also important. Since propagating tides are rotationally influenced internal gravity waves of long period, some estimate of this time may be obtained from the elementary theory of internal gravity waves (Hines, 1960). For long periods

$$C_G = \text{group velocity} \approx \sigma_g L_z^2 / 2\pi L_H, \tag{190}$$

where $\sigma_g = $ Brunt-Väisälä frequency, $L_H = $ horizontal wavelength.

From our earlier discussion of Hough Functions and equivalent depths, we know that for a given choice of σ and s, decreasing positive equivalent depths lead to decreasing values of both L_z and L_H; it turns out that they also lead to decreasing C_G. Let us call the time it takes a wave to traverse a region of depth D, the residence time.

$$\tau_{res.} \approx D / C_G. \tag{191}$$

Now for large enough D, $\tau_{res.} > \tau_{tide}/2\pi$. Thus, it is possible that although $\tau_{diss.} > \tau_{tide}/2\pi$, $\tau_{diss.} < \tau_{res.}$ – in which case dissipation will still be important. It should also be noted that for $\tau_{diss.}$, σ, and s all fixed, $\tau_{res.}$ will be inversely proportional to L_z; hence modes with shorter vertical wavelengths will be preferentially damped – quite apart from any dependence on L_z which $\tau_{diss.}$ might have.

Having crudely discussed the general conditions under which a dissipative process will be of importance, we now turn to specific processes.*

3.6B.1. *Infrared Cooling*

If we assume that on the average there is a balance between infrared flux divergence and other heating and cooling processes, then any perturbation in temperature will lead to an increase or decrease in infrared cooling which will act so as to restore the original temperature. Thus, infrared cooling will tend to damp the temperature component of tidal oscillations – and, in consequence, the tidal oscillation as a whole.

Now, the treatment of infrared radiative transfer is, itself, a matter of considerable complexity (Goody, 1964; Kondratyev, 1960). However, Rodgers and Walshaw (1966) have shown that, over most of the atmosphere, the infrared flux divergence is dominated by 'cooling to space'. That is, it depends primarily on the temperature at the level in question and the amount of infrared active gas between that level and space. This cooling can be fairly well approximated by Newtonian cooling (Lindzen and Goody, 1965; Leovy, 1964); in the present context this can be written:

Infrared cooling rate due to perturbation in temperature

$$\approx a(z, \theta, \phi) \, \delta T, \tag{192}$$

where $a = $ rate coefficient for Newtonian cooling. Lindzen and McKenzie (1967) (see also Lindzen, 1968a) have shown that when a depends only on z, then the inclusion of Newtonian cooling in traditional tidal theory involves only a slight additional complication. Equation (13) becomes

$$\frac{DT}{Dt} = \frac{gH}{\varrho_0} \frac{\gamma - 1}{R} \frac{D\varrho}{Dt} + \frac{\gamma - 1}{R} J - a(z) \delta T. \tag{193}$$

---

* The approximate adiabaticity of tides near the ground was empirically demonstrated by Chapman (1932a).

Equation (16) becomes

$$G = -\frac{1}{\gamma p_0}\frac{Dp}{Dt} = \left(1 + \frac{a}{i\sigma\gamma}\right)^{-1}\left\{\left(1 + \frac{a}{i\sigma\gamma}\right)\chi - \frac{\kappa J}{gH} - \frac{a}{i\sigma\gamma}\frac{w}{H}\frac{dH}{dz}\right\}.$$ (194)

Equation (26) remains the same, while Equation (27) becomes

$$\frac{d^2 L_n^{\sigma,s}}{dx^2} - \left\{1 + \frac{a}{i\sigma\gamma}\left(1 + \frac{a}{i\sigma\gamma}\right)\frac{1}{H}\frac{dH}{dx}\right\}\frac{dL_n^{\sigma,s}}{dx}$$

$$+ \left\{\frac{1}{h_n^{\sigma,s}}\left(\frac{dH}{dx} + \kappa H\right) + \frac{a}{i\sigma\gamma}\frac{1}{H}\frac{dH}{dx}\right\}\left(1 + \frac{a}{i\sigma\gamma}\right)^{-1} L_n^{\sigma,s}$$

$$= \left(1 + \frac{a}{i\sigma\gamma}\right)^{-1}\frac{\kappa}{\gamma g h_n^{\sigma,s}} J_n^{\sigma,s}$$

$$- \left(\frac{a}{i\sigma\gamma}\right)\left(1 + \frac{a}{i\sigma\gamma}\right)^{-1}\left(\frac{1}{H}\frac{dH}{dx}\right)\left(\frac{i\sigma}{\gamma g h_n^{\sigma,s}}\right)\Omega_n^{\sigma,s}.$$ (195)

Using (29) to define $y_n^{\sigma,s}$, one finds that Equations (35) and (38)–(44) remain the same; (37) becomes

$$\delta T_n^{\sigma,s} = \frac{1}{R}\left(1 + \frac{a}{i\sigma\gamma}\right)^{-1}\left\{\frac{\Omega_n^{\sigma,s}}{H}\frac{dH}{dx} - \frac{\gamma g h_n^{\sigma,s}}{i\sigma}e^{x/2}\left[\frac{\kappa H}{h_n^{\sigma,s}}\right.\right.$$

$$\left.\left. + \frac{1}{H}\frac{dH}{dx}\left(\frac{d}{dx} + \frac{H}{h_n^{\sigma,s}} - \frac{1}{2}\right)\right]y_n^{\sigma,s} + \frac{\kappa J_n^{\sigma,s}}{i\sigma}\right\}.$$ (196)

The methods of Section 3.3.B can be applied to (195) as easily as to (27) (or (30)). The ease with which Newtonian cooling can be handled permits incorporating some of the most important features of infrared cooling. Also Newtonian cooling, as a dissipative process, is in many ways representative of other dissipative processes with comparable time scales. Its inclusion, therefore, allows some heuristic inferences to be made on the general role of dissipation.

We return to the last matter explicitly in our discussion of the role of molecular diffusion. Turning to infrared cooling we show in Figure 3.24 a distribution of $a(z)$ due to the CO_2 15 μ band used by Lindzen and Goody (1965). Lindzen and Goody also showed that the effective cooling rate coefficient was greatly enhanced by interaction with ozone photochemistry. The enhanced cooling rate coefficient is also shown in Figure 3.24. No claim is made for the accuracy of our choice of $a(z)$ other than that it represents the correct order of magnitude.

The effect of Newtonian cooling on the solar diurnal thermal tide is discussed by Lindzen (1968a). The effect on all diurnal fields (including surface pressure) below 10 km is negligible. Some idea of the effect on upper air fields may be got from Figure 3.25 where we show the amplitude of the calculated diurnal temperature oscillation as a function of altitude in the absence of Newtonian cooling, and for the two distributions of $a(z)$ in Figure 3.24. Radiation reduces amplitudes by as much as 20%; it also selectively damps the modes of shortest wavelength – as might be expected from our introductory remarks on dissipation.

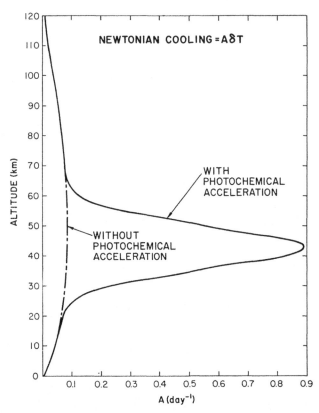

Fig. 3.24. The altitude distribution of the cooling rate coefficient with and without photochemical acceleration. In the absence of photochemical acceleration, it is primarily due to infrared emission by CO_2. After Lindzen (1968a).

The inclusion of the distributions of $a(z)$ shown in Figure 3.24 turns out to have a negligible influence on calculations of the solar semidiurnal thermal tide. On the other hand, Newtonian cooling appears to have a profound effect on the lunar semidiurnal tide. This may be seen in Table 3.8, where we give the calculated amplitude of the lunar semidiurnal surface pressure oscillation for various distributions of T_0 with and without Newtonian cooling. Newtonian cooling sharply reduces sensitivity to variations in T_0. The difference in the effect of Newtonian cooling on the semidiurnal thermal tide and on the lunar semidiurnal tide is easily understood in terms of the introduction to this section, and the discussion in Section 3.5.C. The lunar semidiurnal tide depends significantly on multiple reflections, while the solar semidiurnal thermal tide does not. Thus the effective depth D of the region traversed by the lunar oscillation may be much greater than that traversed by the solar oscillation, and from (191), $\tau_{res.}$ (and hence the effects of dissipation) will then be much greater for the lunar than for the solar oscillation. This general discussion is supported by the fact that the lunar semidiurnal tide in an isothermal atmosphere – where there are no internal reflections – is negligibly affected by infrared cooling. One

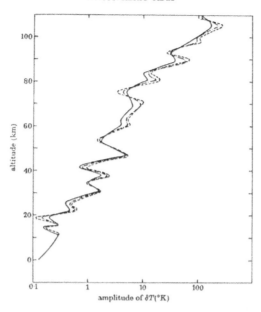

Fig. 3.25. Amplitude of the diurnal temperature oscillation over the equator for different distributions of the cooling rate coefficient: ——, standard; –·–, without photochemical; ————, zero.
After Lindzen (1968a).

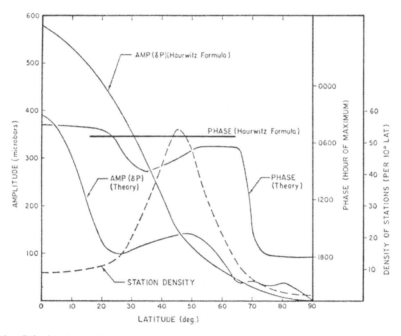

Fig. 3.26. Calculated amplitude and phase of the solar diurnal surface pressure oscillation. These
quantities are also shown as derived from the Haurwitz (1965) empirical formula. The distribution of
stations on which his formula is based are shown. After Lindzen (1967a).

TABLE 3.8

Effect of Newtonian cooling on the lunar semidiurnal surface pressure oscillation for different choices of $T_0(z)$

Temperature model	Newtonian cooling model (see Figure 3.24)	$|\delta p|$	Phase (deg)
Equatorial standard	A [a]	373 μb	320
Equatorial standard	B [b]	157 μb	19
Isothermal ($T_0 = 260°$)	A	60 μb	67
Isothermal ($T_0 = 260°$)	B	62 μb	69
ARDC	A	4.5 μb	46
ARDC	B	39 μb	28
ARDC (modified to have short cold tropopause)	A	15 μb	287
ARDC (modified to have short cold tropopause)	B	44 μb	9

[a] A: no radiative cooling.
[b] B: photochemically accelerated radiative cooling.

expects that the sensitivity of the lunar semidiurnal tide to T_0 will be further reduced with the inclusion of all dissipative mechanisms – not just infrared cooling.

3.6B.2. *Molecular Viscosity and Conductivity*

The inclusion in tidal theory of molecular viscosity and conductivity (and turbulence insofar as it can be described in terms of viscosities and conductivities) is far more difficult than the inclusion of Newtonian cooling. The order of the differential equations is raised and the altitude and latitude dependence of the solutions ceases to be separable (Nunn, 1967). According to Nunn (1967) the lack of separability is due to the existence of coriolis terms in the equations of motion. Thus far no thorough study has been made of the role of viscosity and conductivity in tidal theory – mainly because of these mathematical difficulties. However, those tidal oscillations that propagate vertically are, in fact, internal gravity waves – influenced somewhat by the earth's rotation. Thus, simpler calculations on the effect of viscosity and/or conductivity on internal gravity waves in a plane, non-rotating atmosphere may, at least, suggest what happens in the case of tides.* That molecular viscosity and conductivity are likely to be important for tides – at least at sufficiently great heights – follows from the increase as $1/\varrho_0$ of both thermometric conductivity and kinematic viscosity (viz. (189)). Thus there must be some altitude above which $\tau_{\text{diss.}}$ due to molecular processes is shorter than other relevant time scales.

The obvious first approach to the role of viscosity and conductivity on internal gravity waves is to consider the case where the viscosity and conductivity are small perturbations on the adiabatic wave, and where the vertical wavelength of the wave is less than the density scale height. One may then substitute the adiabatic solutions

---

* Nunn (1967), in studying the effects of viscosity and conductivity on the semidiurnal and terdiurnal thermal tides, omitted the coriolis terms in the region where these molecular processes are important. His equivalent depths, however, were taken from the classical theory, which includes coriolis effects.

into the heat and momentum diffusion terms. This approach, taken by Pitteway and Hines (1963), predicts the decay of waves in the direction of energy propagation – thus supporting the use of the radiation condition in the adiabatic theory. Similar results were obtained in a more complete investigation by Midgley and Liemohn (1966). Neither of the above mentioned studies took complete account of the effect of the fact that kinematic viscosity and thermometric conductivity increase as $1/\varrho_0$. Yanowitch (1967) studied this by considering internal gravity waves in a plane, non-rotating, infinite, exponentially stratified, incompressible, non-conducting fluid where kinematic viscosity is finite and proportional to $1/\varrho_0$. He assumed that at the bottom of his fluid $\tau_{\mathrm{visc.}}$ was much longer than the wave period; the excitation, moreover, was assumed to exist at the lower boundary. Because viscosity was small in the lower portion of the fluid the solutions were wavelike in z there – just as in the adiabatic case. However, the radiation condition did not apply. In addition to an upward propagating (downward phase speed) component of amplitude A, there was a reflected component of amplitude $A \exp(-2\pi^2 H/L_{\mathrm{z}})$, where H is the fluid's scale height and L_{z} is the vertical wavelength the wave would have in the absence of viscosity. The phase of the reflected wave is a complicated function of wave period, molecular viscosity, and horizontal wavelength. The reflection results from the fact that there must exist some level above which viscosity is important, and in the neighborhood of that level viscosity is a significant inhomogeneity in the fluid. Reflection is only important when $L_{\mathrm{z}} \gtrsim 2\pi^2 H$. This might be the case for the main semidiurnal modes. For the short vertical wavelength propagating diurnal modes, the radiation condition appears to be essentially correct. In addition to reflecting waves, viscosity eventually prevented the continued growth of wave amplitudes as $1/\sqrt{\varrho_0}$. For waves with short vertical wavelengths viscosity even produced sharp attenuations of wave amplitudes above some level. The maximum wave amplitudes are found below the level where $\tau_{\mathrm{visc.}} \approx \tau_{\mathrm{wave}}/2\pi$, and the maximum amplitude is the amplitude the wave would have reached, in the absence of viscosity, some distance below the level at which the maximum occurs. Above the level where $\tau_{\mathrm{visc.}} \approx \tau_{\mathrm{wave}}/2\pi$, the wave's phase rapidly becomes independent of height.

Yanowitch's results, though interesting, have uncertain relevancy to atmospheric tides and thermal tides. Some uncertainty has been removed by Lindzen (1968b). He considered a fluid subject only to the approximation described in Section 3.2. The assumption of no dissipation, however, was relaxed in order to include Newtonian cooling – the cooling rate, $a(z)$, being represented as follows:

$$a(z)/\sigma\gamma = \varepsilon e^x, \tag{197}$$

where $\varepsilon \ll 1$; (197) simulates both the decrease of $\tau_{\mathrm{diss.}}$ as $1/\varrho_0$, and the smallness of dissipation in the lower atmosphere. The results obtained using (197) were exactly the same as Yanowitch's as far as reflection is concerned. More surprisingly, the vertical distributions of wave velocity fields were also very similar. This strongly suggests that most of Yanowitch's results depended neither on his assumptions nor on the details of his dissipative mechanism, but on the fact that the dissipation rate was inversely proportional to density. This further suggests that Lindzen's simple calculations might be used to estimate the effects on tides of dissipative processes, like molecular viscosity and heat conduction, which grow in importance as $1/\varrho_0$.

Let x_0 be that height at which the dissipative terms are comparable to local time derivatives (in terms of Equation (197), $\varepsilon e^{x_0} = 1$). Then, for the gravest, propagating diurnal mode ($n = 1$, viz. Section 3.5.B), $2\pi^2 H/L \approx 2$, and x_0 corresponds to $z \approx 110$ km. Lindzen's calculations suggest that above x_0, phase changes with height disappear. Maximum amplitudes for wind occur about 0.3 scale heights below 110 km although amplitudes decrease only 2% above this altitude. The maximum amplitude is that which the adiabatic solution would have reached 1.6 scale heights below x_0 (i.e., $z \approx 95$ km; viz. Figure 3.13). The above general structure appears to correspond to measured winds in the ionosphere (R. J. Reed, private communication). Hines (1966), however, reports decay above 100 km (cf. 2S.7C). This is probably due to the rapid increase of H with height at these levels – leading to values of $2\pi^2 H/L \gg 2$.

3.6B.3. *Ion Drag and Thermal Tides in the Ionosphere*

At altitudes above 100 km the interactions of the ionized and neutral components of the atmosphere assume increasing importance. The complexity of these interactions has thus far prevented their inclusion in a complete tidal theory. However, isolated features of the interaction have been considered. One such feature is the dynamo effect. Briefly, tidal motions will blow charged particles across the earth's magnetic field lines, thus setting up electric fields. In the region 100–150 km, the atmospheric conductivity is high due to the differential mobility of ions and electrons (Chapman, 1956), and these electric fields produce currents which in turn produce tidal perturbations in the magnetic field, which can be measured at the ground (Chapman and Bartels, 1940; Chapman, 1961). There are a considerable number of theoretical studies of atmospheric dynamo theory (Chapman, 1919b; Baker and Martyn, 1953; Kato, 1956; Maeda, 1955; and others). However, they all assume very simple models for tidal wind fields; none provide for the structured tidal fields described in this review.

In the region of high conductivity, ions are readily advected by wind; hence the ions may be presumed to exert little drag on the neutral tidal winds. However, at greater altitudes, where the ionic gyro-frequencies are much greater than their collision frequencies with neutral molecules, the ions are firmly fastened to magnetic field lines (except insofar as they are moved by electric fields conducted up magnetic field lines (Martyn, 1955), and, therefore, exert a drag on neutral air motions across magnetic field lines. In simple cases, this drag takes the form of a Rayleigh friction; i.e., a term of the form $-Du$ should be added to the right hand side of Equation (8) for example. For the component of horizontal flow perpendicular to magnetic field lines near the magnetic equator, Lindzen (1967d) found

$$D \approx 0.5 \times 10^{-9} N_i \sec^{-1}, \tag{198}$$

where N_i = ion number density. In the F-region N_i can reach 10^6 cm^{-3}, in which case

$$D \approx 0.5 \times 10^3 \sec^{-1} \approx 6.86\, 2\pi/1 \text{ day}. \tag{199}$$

Such a value for D clearly establishes ion drag as an important influence on motions of tidal periods at F-region altitudes. Unfortunately, the inclusion of ion drag in tidal calculations of the sort described here has not been achieved – in some part because of the following complications: (a) the inclusion of ion drag (like the inclusion of viscosity) leads to non-separable equations; (b) the region where ion drag is likely to be important is also a region where molecular viscosity and conductivity are likely to be important; (c) N_i has a huge daily variation; (d) thermotidal oscillations above 150 km appear to be excited not only by the upward propagation of thermotidal oscillations below the ionosphere, but perhaps to a greater degree by heat generated by the absorption of solar ultraviolet radiation by O_2 near 120 km – the heat being conducted upwards by molecular collision (Harris and Priester, 1965); (e) electric fields, generated by tidal motions in the dynamo region, can induce tidal motions in the ions aloft. These ions will 'drag' neutral molecules along, thus inducing tidal motions while damping deviations from these motions. Such electrically driven motions can amount to as much as 40 m/sec (Martyn, 1955).

The above complications explain why tidal calculations are difficult to perform for the thermosphere; they do not imply that thermotidal fields are unimportant in the thermosphere. Indeed satellite drag measurements show immense daily variations in thermospheric density (Jacchia, 1963; King-Hele and Walker 1961; and others). Taking into account ion drag and viscosity, it has been shown that, consistent with the above mentioned density variations, there should be daily variations in horizontal wind with amplitudes of about 80–200 m/sec (Geisler, 1966; Lindzen, 1967d; King and Kohl, 1965; and others).

3.6C. NON-LINEAR EFFECTS

One of the assumptions introduced in Section 3.2 was that quadratic and higher order terms in tidal fields arising from nonlinear terms in the equations of motion could be ignored. The resulting equations were linear, and hence, the amplitudes calculated for tidal fields were proportional to the amplitude of the tidal or thermotidal excitation. A necessary condition, therefore, for the validity of linearization was that the excitation be sufficiently small. As pointed out, this was not enough. Tidal modes with sufficiently small positive equivalent depths are, in fact, vertically propagating internal gravity waves, and the fields associated with such modes (namely u, v, w, δT, $\delta p/p_0$, $d\varrho/\varrho_0$) grow – in the absence of dissipation – as $1/\sqrt{p_0}$. Thus, if we neglect dissipation, then for any excitation, however small, there will exist some height above which linearization breaks down. This problem was noticed by Pekeris (1951), and initial attempts were made at setting up a formalism for studying nonlinear effects (Pekeris and Alterman, 1959). However, no results were obtained. One problem is that if we attempt to study nonlinear effects by means of straightforward amplitude expansions then first order (linear) fields will grow with altitude as $1/\sqrt{p_0}$, second order fields will grow as $1/p_0$, etc. There will always be some height above which the expansion fails to converge.

The solution to the above problem may lie in the inclusion of dissipation. As we have seen (Section 3.6.B.2) one effect of the increasing importance of molecular viscosity (and, presumably conductivity) with height is to eliminate the $(\sqrt{1/p_0})$ growth above some altitude. Thus amplitude expansions based on solutions to the viscous, conducting equations might, in fact, converge uniformly. Unfortunately, such a computational procedure seems to be beyond present capabilities.

More encouragingly, the effect of dissipation may prevent fields from growing so large as to make nonlinear terms very important. The preliminary results in Section 3.6.B.2 suggest this. Even if the quadratic and higher order terms are not of preponderant importance they may have important effects. Not only will they cause distortions of the linearized solutions, but – in the presence of dissipation – they can produce steady transports of heat and momentum (Lindzen, 1967e). Detailed investigations of these possibilities remain to be explored.

Should nonlinearities be small, it still appears possible that tidal fields (especially for the short vertical wavelength propagating diurnal modes) can become unstable at high altitudes. Propagating tidal modes, and internal gravity waves in general, appear capable of wrinkling surfaces of constant potential temperature in such a manner as to produce static instability (Lindzen, 1968a; Hodges, 1968); they can also produce high shears. Tides and thermal tides might, therefore, be responsible for the generation of turbulence at high altitudes. Lindzen (1968a) suggests that this will be the case for the solar diurnal thermal tide near 90 km over the equator. Thermal tides, as a result, may cause the dynamic mixing of the upper atmosphere and, in consequence, help to determine the height of the turbopause.

3.6D. Neglect of Mean Winds and Horizontal Temperature Gradients

In the atmosphere, mean winds and horizontal temperature gradients accompany each other and are approximately related by the thermal wind equation

$$\frac{\partial V_0}{\partial z} = \frac{g}{2\omega\cos\theta} \frac{1}{T_0} \frac{1}{a}\left(\frac{\partial T_0}{\partial\theta}\right)_{p\ \mathrm{const.}}. \tag{200}$$

(Eliassen and Kleinschmidt, 1957). Thus, in principle, one cannot deal separately with the effects of mean winds and of horizontal temperature gradients (or, relatedly, mean horizontal gradients of pressure and density). Most treatments of this matter are consistent in this respect; however, emphasis tends to be on one aspect or the other. Thus Haurwitz (1957) and Siebert (1957) emphasized the effects of meridional temperature gradients, while Chiu (1953), Sawada (1966) and Dikii (1967) emphasized the effects of wind. All the above mentioned studies suggest only small effects; however, these studies are also restricted to particular tidal modes and atmospheric models.

Given our current capabilities, a full treatment of tides and thermal tides in atmospheres with arbitrary distributions of wind and temperature seems unlikely. However, some heuristic comments may prove useful. In these comments we will, artificially, consider wind and temperature effects separately.

Some idea of whether latitude variations in T_0 may be important can be obtained by seeing if small changes in $T_0(z)$ make much difference in the calculations described in Section 3.5. The results displayed in Figures 3.7, 3.19, and 3.21 suggest that such changes can indeed be significant; however, this by no means implies that latitude temperature gradients *per se* will be important. Intuitively, it seems that for tidal modes which smoothly span all latitudes (like $\Theta_2^{2\omega,\,2}$, $\Theta_{-2}^{\omega,\,1}$; viz. Figures 3.3 and 3.10) latitude variations in T_0 will not matter much since these modes presumably 'see' latitude averages of T_0. Higher order modes which oscillate with latitude are, one suspects, more sensitive to latitude variations in T_0.

Insofar as tidal modes can be considered internal gravity waves, the importance of wind is given by the ratio of the wind speed to the tidal horizontal phase speed (Booker and Bretherton, 1967). For migrating tides and thermal tides, the zonal phase speeds are approximately given by $\omega a \sin\theta$, where $\omega a = (2\pi/1 \text{ day}) \times 6400 \text{ km} = = 466 \text{ m/sec}$. Since mean winds are generally much smaller than this value, we expect that their effect on tides will be small (at least away from the poles where $\sin\theta = 0$).

3.6E. ADDITIONAL REMARKS

We have, in this section, discussed, from a theoretical point of view, the approximations involved in calculations described in Sections 3.2–3.5. In addition, we must not lose sight of the uncertainties in the precise specification of thermal excitations. The question remains, as to how well the approximate calculations describe the observations. Several aspects of this question have already been answered. The approximate theory does quite well in predicting solar thermotidal oscillations in surface pressure. The theoretical description of lunar tidal oscillations, on the other hand, clearly requires some consideration of dissipative processes.

In the next section, we give a detailed comparison of theoretical results with available data. It will be seen that the approximate theory is quite successful up to at least 95 km for the major thermal tides.

3.7. Comparison of Theory with Data

It should be clear from this chapter that tidal theory is not yet able confidently to predict the details of seasonal variations, local effects, the behavior of tides in the thermosphere, etc. Significantly, however, current tidal theory has been able to predict the average (or typical) structure of the main migrating tides and thermal tides of the atmosphere below 100 km with remarkable quantitative accuracy. Beginning with surface pressure oscillations, we have already noted that present thermotidal theory predicts a solar semidiurnal oscillation almost identical to observations with respect to both amplitude and distribution, and that the prediction is insensitive to details of the atmospheric structure. There is a discrepancy: inviscid tidal theory predicts a semidiurnal pressure maximum near 0900 local time, while the observed maximum occurs between 0930 and 1000 local time. There is some reason to believe that this phase error in the theory is due to the neglect of the influence of the turbulent surface boundary layer. Turning to $S_1(p)$ we find that our current theory predicts oscillations whose amplitude at the equator is about two thirds of the value obtained from data by Haurwitz (1965). The correspondence of theory with Haurwitz's empirical distribution for the migrating part of $S_1(p)$ at other latitudes is shown in Figure 3.26. It is important to note that the agreement is best where the station density is greatest, because, where the station density is low, the empirical determination of the *migrating* solar diurnal oscillation is likely to be inaccurate. This is because the solar diurnal oscillation at a given station consists in comparable contributions from local and migrating parts, and the separation of the two requires data from a good many stations in each latitude belt.

Although Butler and Small (1963) have accounted for most of the observed features of $S_3(p)$ at the ground, little theoretical attention has been given to $S_4(p)$ or to the non-migrating oscillations ... at least when compared with the work on the migrating parts of $S_1(p)$ and $S_2(p)$ at the ground. Similarly, little work has been done on the seasonal dependence of $S(p)$.

For $L_2(p)$ at the ground, we saw in Section 3.5.C that adiabatic tidal theory and annually averaged observations agree only for certain assumed atmospheric tempera-ture structures (most notably isothermal atmospheres). In section 3.6.B we saw that the inclusion of dissipative processes in the theory sharply reduced the temperature dependence of the lunar semidiurnal response – bringing theoretical calculations for all temperature profiles closer to annually averaged observations. Even with dissipation there remains a significant dependence of $L_2(p)$ on T_0. This sensitivity undoubtedly plays a role in the seasonal variability of $L_2(p)$; however, rigorous theoretical treat-ment of this matter must await our ability to deal with realistic latitude variations in T_0.

Above the ground our data refer (with a few exceptions) to solar oscillations. A comparison of the data analysis for $S_2(u)$ in Table 2S.9 with the theoretical calculations displayed in Figures 3.8 and 3.9 show rough agreement of amplitudes and phases up to about 30 mb. The theoretical prediction of a $180°$ phase shift near 25 mb does not appear to be confirmed by the data. Rocket data do suggest a phase shift near 50 km (viz. Figure 2S.15). However, the full resolution of this discrepancy remains to be found. Comparing Figures 3.13 and 3.14 with Table 2S.9 we see that there is rough agreement between theory and observation for $S_1(u)$ above 200 mb. Because non-migrating components contribute significantly to $S_1(u)$ below 15 mb (viz. Section 2S.7A), we should not expect detailed agreement. Wallace and Hartranft (1969) show that the relative importance of non-migrating components is much diminished above 15 mb. Thus we might expect better agreement at altitudes above about 30 km. Figures 2S.13, 2S.14, 2S.16 and 2S.17 show the results of both data analyses and theoretical calculations for the solar diurnal oscillation in meridional wind; the agreement is remarkable. Moreover, Reed *et al.* (1969) have pointed out that the few systematic discrepancies shown in Figures 2S.16 and 2S.17 could be eliminated if the diurnal excitation as given by Equations (168) and (169) consisted in greater con-tributions from the $\Theta_{-2}^{\omega,\,1}$ mode and smaller contributions from the $\Theta_1^{\omega,\,1}$ mode. Such a change is entirely within the realm of possibility. Our expression for $g_{O_3}^{\omega,\,1}$ (Equation (168)) is based on the work of Leovy (1964), which neglected the existence of an ozone maximum near 60 °N and S. The effect of such a maximum on the heating due to ozone insolation absorption is likely to be of the form suggested by Reed, Oard and Sieminski (1969).

Going to higher altitudes, and comparing Figures 2S.18 and 2S.19 with Figures 3.8, 3.9, 3.14 and 3.15, we find that theoretical calculations for diurnal and semidiurnal wind oscillations agree excellently in amplitude with radio meteor data for the region 80–100 km. The data show marked variations of phase with season. The theory goes some way towards explaining the variation. The diurnal and semidiurnal oscillations above 80 km are simply gravity-type waves propagating upwards from the tropo-sphere and stratosphere. Seasonal variations in mean winds and temperatures cause changes in the index of refraction of the atmosphere through which these waves travel, and their phase must also change on arriving in the region 80–100 km. One may presume that if detailed vertical profiles of thermotidal winds were available they would look very similar (except for vertical displacements of phase) in each season. One should note that the averaging thickness of 20 km is very close to the wave-length of the main propagating diurnal mode. Hence, phase shifts will also produce spurious amplitude variations.

Wind data inferred from dynamo calculations are also subject to problems of averaging. These problems, however, are unlikely to be sufficient to explain the disparity between the winds shown in Figure 2S.20 and theoretical calculations. That a disparity should exist is not surprising, because current theory still neglects viscosity, conductivity, and hydromagnetic processes – all potentially important in the dynamo region.

Figure 2S.21 shows the inadequacy of the present theory for the thermosphere most clearly. Agreement between theory and observation for the diurnal oscillation is excellent up to 105 km. However, the present theory fails to predict the observed decay of amplitude above 105 km.

There are certainly many important problems, both theoretical and observational, remaining in the area of atmospheric tides and thermal tides. However, the progress of the last twenty years suggests that we may face these problems with greater con-fidence than was previously possible. Such confidence must, of course, be muted by the surprises that have so long dominated the history of this field.

Acknowledgments

R. S. Lindzen wishes to acknowledge support of the National Science Foundation under Grant No. GA-1622. We wish gratefully to acknowledge the help of Dr. Bernhard Haurwitz in the preparation of this review.

List of Symbols for Chapter 3

In view of the large number of symbols introduced in Chapter 3 we have prepared the following list of symbols giving meanings and pages on which the symbol is first introduced. The list is not complete; many symbols used on only a few pages are omitted. In addition the following superscript notation is used:

(1) f^+ means that f is also used with the subscript zero referring to the value of f in the basic state (undisturbed by tidal or thermotidal oscillations).

(2) f^* means that f appears as $f^{\sigma, s}$ where σ refers to the frequency of an oscillation and s to the zonal wave number.

(3) f^{**} means that f also appears as $f_n^{\sigma, s}$ where n refers to the nth term in a Hough function expansion.

(4) f^{***} means that f appears as $f_{n, m}^{\sigma, s}$, the additional symbol m refers to the mth term in an associated Legendre polynomial expansion of the nth Hough function.

Symbol	Meaning	Page
a	earth's radius	107
a	rate coefficient for Newtonian cooling	160
C_p, C_v	heat capacities at constant pressure and constant volume	109
C^{***}	coefficient in the expansion of Hough functions in terms of associated Legendre polynomials	114
$\bar{C}^{**}$	coefficient in the expansion of normalized Hough functions in terms of normalized associated Legendre polynomials	120
c_G	vertical component of group velocity	160
D	earth-moon or earth-sun distance	121
D	ion drag coefficient	166
f	$\sigma/2\omega$	109
F	operator associated with Laplace's tidal equation	110
f_G^*	vertical distribution of τ_G	126
g	acceleration of gravity	107
$G^{*, **}$	$-\dfrac{1}{\gamma p_0}\dfrac{Dp}{Dt}$	109
$g_G^{*, **}$	latitude distribution of τ_G	126
h_θ, h_ϕ, h_r	metric factors for spherical polar coordinates	107
H	scale height	108
h^{**}	separation constant in separating equations for G's altitude and colatitude dependence; equivalent depth	110
$J^{*, **}$	thermotidal heating per unit time per unit mass	108
$J_G^{*, **}$	part of J due to absorption of radiation by gas, G	125
$K_{1m} + K_{1s}$	diurnal luni-solar gravitational potential	123
$K_{2m} + K_{2s}$	semidiurnal luni-solar gravitational potential	123
K	eddy thermal conductivity	124
L^{**}	vertical variation of G	110
L_H	horizontal length scale for tidal field	159
L_z	vertical wavelength scale for tidal field	159

M	mass of moon or sun	121
M_2	semidiurnal term in lunar gravitational potential	123
N_2	large lunar elliptic semidiurnal term in gravitational potential	123
N_i	ion number density	166
O_1	diurnal term in lunar gravitational potential	123
$p^{+,*,**}$	pressure	106
N_n^s	Neumann form of associated Legendre polynomial (as defined and denoted by P_n^s by Whittaker and Watson, 1927, p. 323)	114
$\bar{P}_{s,n}$	fully normalized associated Legendre polynomial (see p. 36)	120
P_1	diurnal term in solar gravitational potential	123
R	gas constant for air	106
r	distance from the earth's center	107
s	zonal wave number	109
S_2	semidiurnal term in solar gravitational potential	123
$T^{+,*,**}$	temperature	106
t	time	108
$u^{+,*,**}$	southward (or northerly) velocity component	108
U^{**}	expansion function for u's colatitude variation	111
$v^{+,*,**}$	eastward (or westerly) velocity component	108
V^{**}	expansion function for v's latitude variation	112
$w^{+,*,**}$	vertical velocity component	
x	height in scale heights	108
$y^{*,**}$	$e^{-x/2}L$	111
z	altitude	107
α_m, β_m	dummy variables used in numerical integration of vertical structure equation	117
γ	$C_\mathrm{p}/C_\mathrm{v}$	108
γ	gravitational constant	121
$\delta(\Delta)$	variation of (Δ) from its value in the basic state	108
δx	grid interval in the numerical integration of the vertical structure equation	117
ε	magnitude of horizontal gradient of surface elevation	157
ε	ratio of Newtonian cooling rate coefficient to $\gamma\sigma$ at the ground when Newtonian cooling rate is presumed to increase as e^x.	165
θ	colatitude	108
Θ^{**}	colatitude variation of G; Hough function	110
$\bar{\Theta}^{**}$	fully normalized Hough function	120
κ	$(\gamma-1)/\gamma = 2/7$	109
λ	vertical wavenumber when wavelength is expressed in scale heights	113
μ	$\cos\theta$	113
ν	molecular viscosity	159
ξ	surface elevation	157

$\varrho^{+,*,**}$	density	106
σ	frequency of oscillation	109
σ_0	$2\pi/$sidereal day	122
σ_1^L	$2\pi/$lunar month	122
σ_1^S	$2\pi/$year	122
σ_2^L	$\sigma_0 - \sigma_1^L$	122
σ_2^S	$\sigma - \sigma_1^s$	122
σ_g	Brunt-Väisälä frequency	160
$\tau^{*,**}$	$\kappa J/i\sigma R$: temperature change that would be produced by J in absence of other processes	126
$\tau_G^{*,**}$	$\kappa J_G/i\sigma R$; see J_G	126
$\tau_{\text{diss.}}$	time scale for dissipation	159
$\tau_{\text{visc.}}$	time scale for viscous dissipation	159
τ_{tide}	tidal period	159
$\tau_{\text{res.}}$	residence time for a wave in a region of specified depth	160
ϕ	east longitude	108
χ	velocity divergence	108
ω	earth's rotation rate	108
$\Omega^{*,**}$	tidal gravitational potential	108

GUIDE TO THE FIGURES AND TABLES

The italic numbers are those of the pages on which the Figure or Table appears. The following numbers are those of the pages on which it is mentioned in the text. The letter is that of the Figure or Table classification; see pp. 176 to 178, where the nature of the Figure or Table is briefly indicated.

FIGURES

TABLES (CHAPTERS 2S, 2L AND 3)

CLASSIFICATION OF THE FIGURES

All the Figures in Chapters 1, 2S and 2L are observational, and all those in Chapter 3 are theoretical, with the following exceptions, (a), (b):

(a) *Geometrical*
   2L.2, relating to solar time and lunar time and phase,
   3.1, relating to the tidal potential.

(b) *Idealized distributions*
   3.2, height and latitude distributions of thermal excitation by H_2O and O_3 absorption,
   3.6 (ARDC *et al.*), 3.18, 20, atmospheric temperature profiles.
   3.24, height distributions of Newtonian cooling rates by radiation from CO_2 with and without acceleration due to ozone photochemistry. In Chapter 2S, Figures 8, 13, 14, 17, and Figure 3.26, combine or compare observation with formulae or theory.
   The Figures not above mentioned belong to the following types.

(c) *Barographs, time graphs*
   1.1: $p(t)$, 5 days, Batavia and Potsdam,
   1.2: $S_2(p)$, Washington, D.C., for one solar day,
   2L.1: $L_2(p)$, Greenwich, for one lunar day,
   2S.12: $u(t)$, 2 days, White Sands, at each of six heights.

(d) *Height distribution graphs for various latitudes; see also (b)*
   2S.13.14: $s_1(u)$, $\sigma_1(u)$ respectively, 30–62 km, 30°N, observational, and (for j, e, d) theoretical, for isothermal atmosphere,
   2S.15: $s_2(u)$, $\sigma_2(u)$, 30–60 km, 30°N, for isothermal atmosphere,
   2S.16: $s_1(u)$, $\sigma_1(u)$, 30–60 km, 61°N, for isothermal atmosphere,
   2S.17: $s_1(u)$, $\sigma_1(u)$, 30–60 km, 20°N, for isothermal atmosphere,
   3.7: $s_2(v)$, ground to 110 km, equator, for different $T(z)$ profiles,
   3.8, 9: $s_2(u)$ and $\sigma_2(u)$ respectively, ground to 115 km, at each of four latitudes, for equatorial type atmosphere (see Fig. 3.6),
   3.22, 23: $s_2(v)$, $l_2(v)$, and $\sigma_2(v)$, $\lambda_2(v)$ respectively, ground to 111 km, equator, ARDC,
   3.19: $s_1(v)$, noon, 35–100 km, 25°N, for four different $T(z)$ profiles,
   3.13, 14: $s_1^1(u)$ and $\sigma_1^1(u)$ respectively, ground to 105 km, for each of six latitudes, for an isothermal atmosphere,
   3.17:   the S_1 part of u at noon and at 6 p.m., ground to 105 km, latitudes 30° and 50°, for an isothermal atmosphere,
   3.25: $s_1(T)$ at the equator from the ground up to 110 km, for an isothermal atmosphere, for Newtonian cooling absent, or at the two rates shown in Fig. 3.24.
   Different heights are considered in (c), 2S.12 and in (g), 2S.21.

(e) *Latitude distribution graphs (ground level, except for 3.15, 16); see also (b), 3.2*
   2S.8: $s_1^1(p)$, compared with two formulae,
   2L.6: $s_2(p)$, $l_2(p)$.
   3.3, 10:   symmetric Hough functions for S_2^2 and S_1^1 respectively,
   3.4, 5:   expansion functions U_n, V_n, for $S_2^2(u)$, and $S_2^2(v)$ respectively,
   3.11, 12:   expansion functions U_n, V_n, for $S_1^1(u)$, and $S_1^1(v)$ respectively,
   3.15, 16: s_1 and σ_1 respectively, for u, v, w and T, at 85 km, isothermal atmosphere,
   3.26: s_1^1 and σ_1^1 respectively, for u, v, w, and T, at the ground, isothermal atmosphere.
   Different latitudes are considered in (d) 3.13, 14, 17 and in (l) 2S.1 and 2L.5, 7.

(f) *Amplitudes of latitude-averaged longitudinal component waves*
   2S.4:   for $S_2(p)$,
   2S.7:   for $S_1(p)$,
   2L.4:   for $L_2(p)$,

(g) *Vectograms*
    2S.21:   dawn values of $S_1(\mathbf{V})$, and dawn and dusk values of the prevailing wind $+ S_2(\mathbf{V})$, at heights from 90 to 130 km.
    2L.10: $\mathbf{V}$ for S_2 and for L_2, Mauritius, also particle paths.

(h) *Maps showing equilines*
    2S.3: $s_2(p)$, $\sigma_2(p)$
    2S.5:   isobars of $W_{2,3}{}^3$, part of $S_2(p)$,
    2S.6: $A_1(p)$, $B_1(p)$,
    2L.3: $l_2(p)$,

(i) *Maps showing wind vectors or vector differences; see also (g)*
    2S.9: $\mathbf{V}_0 - \mathbf{V}_{12}$ and $\mathbf{V}_3 - \mathbf{V}_{15}$ (the subscripts indicate hours) at 700 mb level, N hemisphere,
    2S.10, 11: $\mathbf{V}_0 - \mathbf{V}_{12}$ at 60 mb and 15 mb levels respectively, N hemisphere.
    2S.20: $\mathbf{V}$ in the E layer, at 15° intervals of longitude (or hours) and 10° intervals of latitude, due to S_1 and to S_2 (from geomagnetic data),

(j) *Harmonic dials*
    1.2: $S_2(p)$ for Washington, D.C.,
    1.3: $L_2(p)$ for Batavia for each of forty years, and the 40-year mean,
    2S.18, 19:   for Jodrell Bank and for Adelaide, Australia, respectively, $S_1(u)$, $S_1(v)$, $S_2(u)$, $S_2(v)$, y, j, e, d, averaged over 80 to 100 km height,
    2L.9: $L_2(u)$, $L_2(v)$, $S_2(u)$, $S_2(v)$ for Mauritius,
    2L.11: $L_2(T)$, Batavia,
    2L.12: $L_2(h_E)$, London, j.
    3.21: $L_2(p)$, at the equator at the ground, for six $T(z)$ profiles.

(k) *Maps showing dial vectors*
    1.4: $S_2(p)$, U.S.A.
    2L.8: $S_2(p)$, $L_2(p)$.

(l) *Dialgrams showing the curves traced out by dial points for a series of times or cases*
    2S.1, 2L.7: monthly, for $S_2(p)$ and for $L_2(p)$ respectively, for stations in four or five latitudes,
    2S.2:   monthly, for $S_2(T)$, mean latitude 55°N, and for the annual and semi-annual components of $S_2(T)$,
    2L.5: $L_2(p)$ as a function of latitude, four dialgrams, for the year y and for the seasons j, e, d.

CLASSIFICATION OF THE TABLES (CHAPTERS 2S, 2L, 3)

(a) *Computed:*
    2S.1:   Calendar of daily season (Sigma) numbers; Bartels (1954).
    2S.11, 12: Relating to the uncertainties in deriving S_1 and S_2 from data covering only part of a day.
    3, 1 to 7: Coefficients in the expansion of various normalized Hough functions $\Theta(L, M)$ in terms of normalized associated Legendre functions; the equivalent depth is given in each case; the periodicity, solar or lunar, diurnal or semidiurnal, is indicated by S_1, S_2 or L_2; Flattery (1967)

Nature of $\Theta(L, M)$	Gravitational, $M > 0$.		Rotational, $M < 0$.
Symmetric:	Tables 3.1: S_2	3.3: S_1	3.4: S_1
Antisymmetric:	Tables 3.2: S_2	3.5: S_1	3.6: S_2
Symmetric:	Tables 3.7: L_2		

(b) *Amplitude and phase of harmonic component oscillations at different latitudes or different heights*
    2S.2, 3: $s_2{}^2$ and $\sigma_2{}^2$ and $s_2{}^0$ and $\sigma_2{}^0$ for the travelling and zonal parts of $S_2(p)$ at various latitudes, according respectively to (i) Simpson (1918), (ii) Haurwitz (1956).

2S.9: s_n and σ_n, for $n = 1$ and $n = 2$ respectively, for the wind components u, v in the Azores, at 29 levels between the ground and 15 mb; Harris, Finger and Teweles (1962).

2L.2: l_2 and λ_2, with probable errors, for 104 stations in different latitudes (y) and 85 stations (j, e, d); Haurwitz and Cowley (1970)

2L.4: l_2 and λ_2, with probable errors, for u, v at 7 stations; Haurwitz and Cowley, 1968, 1969; Chapman, 1948

(c) *Amplitude and phase (observational* or theoretical+) of the main spherical harmonic components in the expression of various solar or lunar daily variations*

2S.4:   For $S_2(p)$, Kertz (1956b).*

2S.5:   For $S_1(p)$, Haurwitz (1965)*

2S.6:   For $S_4(p)$, y, j, e, d, Kertz (1956c)*

2S.7:   For $S_1(T)$, Haurwitz and Möller (1955)*, Kertz (1957)+, Siebert (1961)+.

2L.3:   For $L_2(p)$; Haurwitz and Cowley (1970)

(d) *Miscellaneous*

2S.8:   Deviation of the wind components u, v at balloon level, U.K., from the daily mean wind, at trihourly intervals, Johnson (1955).

2S.10:   The number of cases in which, at Jodrell Bank, U.K. and Adelaide, Australia, the mean wind component u, v, averaged over the height range 80 to 100 km, exceeds or is less than the corresponding amplitude $s_n(u)$, $s_n(v)$, for $n = 1, 2$. Haurwitz (1964).

2L.1:   Historical.

3.8:   Showing, for Newtonian cooling of two types, the effect on $L_2(p)$ according to various standard atmospheres, Lindzen (1968a).

REFERENCES

Meteorological and Geoastrophysical Abstracts **14** (1963), 3958–4019, 'Lunar influences on atmospheric and geophysical phenomena', by Wilhelm Nupon and Geza Thuronyi, gives 313 abstracts of date from 1825 to 1963 (but not complete for this period). It was prepared in conjunction with the IAGA/IAMAP Lunar Committee. Extensive bibliographies of such lunar tidal papers are given also by Chapman (1951) and Chapman and Westfold (1956).

Airy, G. B.; 1877, *Greenwich Meteorological Reductions, 1854–1873, Barometer*, London (See pp. 10, 14, 30).

Angot, A.: 1887, 'Étude sur la marche diurne du baromètre', *Ann. Bur. meteorol. France.*

Appleton, E. V. and Weekes, K.: 1939, 'On Lunar tides in the upper atmosphere', *Proc. Roy. Soc.* **A171**, 171–187.

Appleton, E. V. and Weekes, K.: 1948, 'On lunar tides in the upper atmosphere', *Proc. Roy. Soc. London*, **A171**, 171–187.

Bacon, Roger: 1859, *Opera* (ed. J. S. Brewer), Rolls Series, London.

Baker, W. G. and Martyn, D. F.: 1953, 'Electric currents in the ionosphere', *Phil. Trans. Roy. Soc. London* **A246**, 281–294.

Bartels, J.: 1927, 'Über die atmosphärischen Gezeiten', *Abh. Preuss. Meteorol. Inst.* **8**, Nr. 9.

Bartels, J.: 1928, 'Gezeitenschwingungen der Atmosphäre', *Handbuch der Experimentalphysik* **25** (*Geophysik* 1), 163–210.

Bartels, J.: 1932a, 'Tides in the atmosphere', *Sci. Monthly* **35**, 110–130.

Bartels, J.: 1932b, 'Statistical methods for research on diurnal variations', *Terr. Magnet. Atmos. Elec* **37**, 291–302.

Bartels, J.: 1938, 'Berechnung der lunaren atmosphärischen Gezeiten aus Terminablesungen am Barometer', *Gerlands Beiträge Geophys.* **54**, 56–75.

Bartels, J.: 1954, 'A table of daily integers, seasonal, solar, lunar and geomagnetic', Sci. Report No. 2 AF 19(604)–503, Geophys. Inst. Univ. Alaska.

Bartels, J.: 1957, 'Gezeitenkräfte', in *Handbuch der Physik* **48**, 734–774, Springer, Berlin.

Bartels, J. and Fanselau, G.: 1937, 'Geophysikalischer Mondalmanach', *Z. Geophys.* **13**, 311–328.

Bartels, J. and Fanselau, G.: 1938a, 'Geophysical lunar almanac', *Terr. Magnet. Atmos. Elec.* **43**, 155–158.

Bartels, J. and Fanselau, G.: 1938b, 'Geophysikalische Mondtafeln 1850–1875', *Geophys. Inst. Potsdam, Abh.* Nr. 2, Springer, Berlin.

Bartels, J. and Horn, W.: 1952, 'Gezeitenkräfte', in Landolt-Börnstein, *Zahlenwerte und Funktionen aus Physik*, usw., 6 Aufl., **3**, 271–283, Springer, Berlin.

Bartels, J. and Johnston, H. F.: 1940, 'Geomagnetic tides in horizontal intensity at Huancayo', *Terr. Magnet. Atmos. Elec.* **45**, 269–308; 485–512.

Bartels, J. and Johnston, H. F.: 1940a, 'Some features of the large geomagnetic tides in the horizontal force at Huancayo', *Trans. Amer. Geophys. Union*, 1940, 273–287.

Bartels, J. and Kertz, W.: 1952, 'Gezeitenartige Schwingungen der Atmosphäre', in Landolt-Börnstein, *Zahlenwerte und Funktionen aus Physik*, usw., 6 Aufl. **3**, 674–685, Springer, Berlin.

Bartels, J., Chapman, S., and Kertz, W.: 1952, 'Gezeitenartige Schwingungen der Atmosphäre', in Landolt-Börnstein, *Zahlenwerte aus Physik, Chemi, Astronomie, Geophysik und Technik* (ed. by J. Bartels and P. ten Bruggencate), vol. 3, Springer, Berlin, pp. 680–682. References giving details of these determinations can be found there.

Belousov, S. L.: 1962, *Tables of normalized associated Legendre-Polynomials*, Pergamon Press, London.

Bergsma, P. A.: 1871, 'Lunar Atmospheric Tide', *Obsns. Magn. Meteor. Obs. Batavia* **1**, 19–25 (contains meteorological observations for 1865 to 1868). Calculations of this type, for individual years, carried out in routine fashion, have been published for the 40 years 1866–1905. See also summary for these years in **28** (for 1905), 102ff., published 1907.

Best, N., Havens, R., and LaGow, H.: 1947, 'Pressure and temperature of the atmosphere up to 120 km', *Phys. Rev.* **71**, 915–916.

Beyers, N. J. and Miers, B. T.: 1965, 'Diurnal temperature change in the atmosphere between 30 and 60 km over White Sands Missile Range', *J. Atmos. Sci.* **22**, 262–266.

Beyers, N. J., Miers, B. T., and Reed, R. J.: 1966, 'Diurnal tidal motions near the stratopause during 48 hours at White Sands Missile Range', *J. Atmos. Sci.* **23**, 325–333.

Bjerknes, J.: 1948, 'Atmospheric tides', *J. Marine Res.*, **7**, 154–162.

Blamont, J. E. and Teitelbaum, H.: 1968, 'La rotation du vecteur vitesse horizontal dans les marées atmosphériques', *Ann. Geophys.* **24**, 287–391.

Booker, J. R. and Bretherton, F. P.: 1967. 'The critical layer for internal gravity waves in a shear flow', *J. Fluid Mech.* **27**, 513–539.

Börnstein, R.: 1891, 'Eine Beziehung zwischen dem Luftdruck und dem Stundenwinkel des Mondes', *Meteorol. Z.* **8**, 161–170.

Bouvard, A.: 1830, 'Mémoire sur les observations météorologiques faites à l'Observatoire Royal de Paris', *Mém. Acad. Roy. Sci. Paris* **7**, 267–341 (read April 1827, published after Laplace's death in 1830).

Brillouin, M.: 1932, 'Les latitudes critiques', *Compt. Rend. Acad. Sci. Paris* **194**, 801–804.

Brooks, C. E. P.: 1917, '*The reduction of temperature observations to mean of 24 hours and the elucidation of the diurnal variation in the continent of Africa*', *Quart. J. Roy. Meteorol. Soc.* **45**, 375–387.

Broun, J. A.: 1874, *Trevandrum Magnetical Observations* **1**.

Bruce, G. H., Peaceman, D. W., Rachford, H. H. Jr., and Rice, J. D.: 1953, *Trans. Am. Inst. Min. Metall. Engnrs.* **198**, 79–92.

Brunt, D.: 1939, *Physical and Dynamical Meteorology*, Cambridge University Press, Cambridge, England.

Butler, S. T. and Small, K. A.: 1963, 'The excitation of atmospheric oscillations', *Proc. Roy. Soc.* **A274**, 91–121.

Casson, L.: 1959, *The Ancient Mariners*, Gollancz, London.

Chambers, C.: 1887, 'On the luni-solar variations of magnetic declination and horizontal force at Bombay, and of declination at Trevandrum', *Phil. Trans. Roy. Soc. London* **A178**, 1–43.

Chapman, S.: 1913, 'On the diurnal variations of the earth's magnetism produced by the moon and sun', *Phil. Trans. Roy. Soc. London* **A213**, 279–321.

Chapman, S.: 1918a, 'The lunar atmospheric tide at Greenwich', *Quart. J. Roy. Meteor. Soc.* **44**, 271–280.

Chapman, S.: 1918b, 'An example of the determination of a minute periodic variation as illustrative of the law of errors', *Monthly Notices Roy. Astron. Soc.* **78**, 635–638.

Chapman, S.: 1919a, 'The solar and lunar diurnal variations of the earth's magnetism', *Phil. Trans. Roy. Soc.*, London **A-218**, 1–118.

Chapman, S.: 1919b, 'The lunar tide in the earth's atmosphere', *Quart. J. Roy. Meteorol. Soc.* **45**, 113–139.

Chapman, S.: 1924a, 'The semi-diurnal oscillation of the atmosphere, *Quart. J. Roy. Meteorol. Soc.* **50**, 165–195.

Chapman, S.: 1924b, 'Lunar atmospheric tide at Mauritius and Tiflis', *Quart. J. Roy. Meteorol. Soc.* **50**, 99–112 (abstract: *Nature* **113**, 326).

Chapman, S.: 1930, 'On the determination of the lunar atmospheric tide', *Z. Geophys.* **6**, 396–420.

Chapman, S.: 1932a, 'The lunar diurnal variation of atmospheric temperature at Batavia, 1866–1928', *Proc. Roy. Soc.*, London **A137**, 1–24.

Chapman, S.: 1932b, 'On the theory of the lunar tidal variation of atmospheric temperature', *Mem. Roy. Meteorol. Soc.* **4**, 35–40.

Chapman, S.: 1936, 'The lunar atmospheric tide at Glasgow', *Proc. Roy. Soc. Edin.* **A56**, 1–5.

Chapman, S.: 1948, 'Some meteorological advances since 1939', Procès-Verbaux des Séances de l'Association de Météorologie, Oslo, 2–24, (Uccle, Belgium 1950).

Chapman, S.: 1951, 'Atmospheric tides and oscillations', in *Compendium of Meteorology*, Boston, pp. 262–274.

Chapman, S.: 1952, 'The calculation of the probable error of determinations of the lunar daily harmonic component variations in geophysical data', *Australian J. Sci. Res.* **A5**, 218–222.

Chapman, S.: 1956, 'The electrical conductivity of the ionosphere: a review', *Nuovo Cim.* (Suppl. **4**) **4**, Serie X, 1385–1412.

Chapman, S.: 1961, 'Regular motions in the ionosphere: electric and magnetic relationships', *Bull. Amer. Meteorol. Soc.* **42**, 85–100.

Chapman, S.: 1967, 'The correction for non-cyclic variation in harmonic analysis', *J. Atmos. Terr. Phys.* **29**, 1625–1627.

Chapman, S.: 1969, 'The lunar and solar semidiurnal variations of barometric pressure at Copenhagen, 1884–1949 (66 years)', *Quart. J. Roy. Meteorol. Soc.* **95**, 381–394.

Chapman, S. and Austin, M.: 1934, 'The lunar atmospheric tide at Buenos Aires 1891–1910', *Quart. J. Roy. Meteorol. Soc.* **60**, 23–28.

Chapman, S. and Bartels, J.: 1940, *Geomagnetism*, Vols. I and II, Clarendon Press, Oxford.

Chapman, S. and Miller, J. C. P.: 1940, 'The statistical determination of lunar daily variations in geomagnetic and meteorological elements', *Monthly Not. Geophys. Suppl.* **4**, 649–669.

Chapman, S. and Hofmeyr, W. L.: 1963, 'The solar and lunar daily variations of barometric pressure at Kimberley (28.7°S, 24.8°E; 1202 meters), 1932–1960', *Pretoria Wea. Bur. J. 'Notos''* **12**, 3–18.

Chapman, S. and Falshaw, E.: 1922, 'The lunar atmospheric tide at Aberdeen, 1869–1919', *Quart. J. Roy. Meteorol. Soc.* **48**, 246–250.

Chapman, S. and Tschu, K. K.: 1948, 'The lunar atmospheric tide at twenty-seven stations widely distributed over the globe', *Proc. Roy. Soc.* **A195**, 310–323.

Chapman, S. and Westfold, K. C.: 1956, 'A comparison of the annual mean solar and lunar atmospheric tides in barometric pressure as regards their world-wide distribution of amplitude and phase', *J. Atmos. Terr. Phys.* **8**, 1–23.

Charney, J. G. and Drazin, P. G.: 1961, 'Propagation of planetary-scale disturbances from the lower into the upper atmosphere', *J. Geophys. Res.* **66**, 83–110.

Charney, J. G. and Eliassen, A.: 1949, 'A numerical method for predicting the perturbations of middle latitude westerlies', *Tellus* **2**, 38–54.

Chiu, W. C.: 1953, 'On the oscillations of the atmosphere', *Arch. Meteorol. Geophys. Biokl.* **A5**, 280–303.

CIRA: 1965, *Cospar International Reference Atmosphere*, North-Holland, Amsterdam.

Craig, R. A.: 1965, *The Upper Atmosphere*, Academic Press, New York.

Darwin, G. H.: 1901, *The Tides*; new edition, 1962, Freeman, San Francisco.

Diehl, W. S.: 1948, Standard Atmospheric Tables and Data, NACA.

Dikii, L. A.: 1965, 'The terrestrial atmosphere as an oscillating system', *Izvestiya, Atmospheric and Oceanic Physics* – English Edition, **1**, 275–286.

Dikii, L. A.: 1967, 'Allowance for mean wind in calculating the frequencies of free atmospheric oscillations', *Izvestiya, Atmospheric and Oceanic Physics* – English Edition, **4**, 583–584.

Doodson, A. T.: 1922, 'The harmonic development of the tide-generating potential', *Proc. Roy. Soc.* **A100**, 305–329.

Duclay, F. and Will, R.: 1960, 'Étude de la variation semi-diurne lunaire de la pression atmosphérique à Tamanrasset', *Comptes Rendus* **251**, 3028–3030.

Duperier, A.: 1946, 'A lunar effect on cosmic rays', *Nature* **157**, 296.

Eckart, C.: 1960, *Hydrodynamics of Oceans and Atmospheres*, Pergamon Press, New York.

Egedal, J.: 1956, 'On the computation of lunar daily variations in geomagnetism: two simple methods', *Publ. Danske Meteorol. Inst.* No. 22.

Eisenlohr, O.: 1843, 'Untersuchungen über das Klima von Paris und über die vom Monde bewirkte atmosphärische Ebbe und Fluth', *Pogg. Ann. Phys. Chemie* **60**, 161–212.

Elford, W. G.: 1959, 'A study of winds between 80 and 100 km in medium latitudes', *Planetary Space Sci.* **1**, 94–101.

Eliassen, A. and Kleinschmidt, E.: 1957, 'Dynamic Meteorology', in *Handbuch der Physik* (S. Flügge, ed). **48**, 1–154, Springer, Berlin.

Eliassen, A. and Palm, E.: 1961, 'On the transfer of energy in stationary mountain waves', *Geofysiske Publ.* **22**, 1–23.

Elliot, C. M.: 1852, 'On the lunar atmospheric tide at Singapore', *Phil. Trans. Roy. Soc. London* **142**, 125–129.

Finger, F. G. and McInturff, R. M.: 1968, 'The diurnal temperature range of the middle stratosphere', *J. Atmos. Sci.* **25**, 1116–1128.

Flattery, T. W.: 1967, *Hough Functions*. Technical Report, No. 21, Dept. of Geophysical Sciences, University of Chicago.

Frost, R.: 1960, 'Pressure variation over Malaya and the resonance theory', *Air Ministry, Scient. Pap.* **4**, 13 pp.

Geisler, J. E.: 1966, 'Atmospheric winds in the middle latitude F-region', *J. Atmos. Terr. Phys.* **28**, 703–721.

Giwa, F. B. A.: 1967, 'Thermal conduction and viscosity and the choice of the upper level boundary condition in the theory of atmospheric oscillations', *Quart. J. Roy. Meteorol. Soc.* **93**, 242–246.

Goldstein, S.: 1938, *Modern Developments in Fluid Dynamics* (2 vols.), Oxford University Press, Oxford.

Golitsyn, G. S.: 1965, 'Damping of small oscillations in the atmosphere', *Izvestiya, Atmospheric and Oceanic Physics* – English Edition, **1**, 82–89.

Golitsyn, G. S. and Dikii, L. A.: 1966, 'Oscillations of planetary atmospheres as a function of the rotational speed of the planet', *Izvestiya, Atmospheric and Oceanic Physics* – English Edition, **2**, 137–142.

Goody, R. M.: 1960, 'The influence of radiative transfer on the propagation of a temperature wave in a stratified diffusing medium', *J. Fluid Mech.* **9**, 445–454.

Goody, R. M.: 1964, *Atmospheric Radiation*, Oxford University Press, London.

Green, J. S. A.: 1965, 'Atmospheric tidal oscillations: an analysis of the mechanics', *Proc. Roy. Soc.* **A288**, 564–574.

Greenhow, J. S. and Neufeld, E. L.: 1961, 'Winds in the upper atmosphere', *Quart. J. Roy. Meteorol. Soc.* **87**, 472–489.

Hann, J. v.: 1889, 'Untersuchungen über die tägliche Oscillation des Barometers', *Denkschr. Akad. Wiss. Wien*, Abt. I, **55**, 49–121.

Hann, J. v.: 1892,' Weitere Untersuchungen über die tägliche Oscillation des Barometers', *Denkschr. Akad. Wiss. Wien* **59**, 297–356.

Hann, J. v.: 1906, 'Der tägliche Gang der Temperatur in der inneren Tropenzone', *Denkschr. Akad. Wiss. Wien* **78**, 249–366.

Hann, J. v.: 1915, in *Lehrbuch der Meteorologic* (Leipzig, C. H. Tauchnitz), 3, Aufl. 1915; 4. Aufl. (with R. Süring), 1926; 5. Aufl. 1938, W. Keller, Leipzig.

Hann, J. v.: 1918a, 'Untersuchungen über die tägliche Oscillation des Barometers. Die dritteltägige (achtstundige) Luftdruckschwankung', *Denkschr. Akad. Wiss. Wien* **95**, 1–64.

Hann, J. v.: 1918b, 'Die jährliche Periode der halbtägigen Luftdruckschwankung', *S.B. Akad. Wiss. Wien*, Abt. IIa, **127**, 263–365.

Hann, J. v.: 1919, 'Die ganztägige (24stundige) Luftdruckschwankung in ihrer Abhängigkeit von der Unterlage (Ozean Bodengestalt)', *Sitzb. Akad. Wiss. Wien*, Abt. IIa, **128**, 379–506. Reference [9] of Chapman, 1951, p. 110 gives corrections.

Harris, I. and Priester, W.: 1965, 'On the dynamical variation of the upper atmosphere', *J. Atmos. Sci.* **22**, 3–10.

Harris, M. F.: 1959, 'Diurnal and semidiurnal variations of wind, pressure and temperature in the troposphere at Washington, D.C.', *J. Geophys. Res.* **64**, 983–995.

Harris, M. F., Finger, F. G., and Teweles, S.: 1962, 'Diurnal variations of wind, pressure, and temperature in the troposphere and stratosphere over the Azores', *J. Atmos. Sci.* **19**, 136–149.

Harris, M. F., Finger, F. G., and Teweles, S.: 1966, 'Frictional and thermal influences in the solar semidiurnal tide', *Mon. Wea. Rev.* **94**, 427–447.

Haurwitz, B.: 1940, 'The motion of atmospheric disturbances on a spherical earth', *J. Mar. Res.* **3**, 254–267.

Haurwitz, B.: 1951, 'The perturbation equations in meteorology', in *Compendium of Meteorology* (T. F. Malone, ed.), American Meteorological Soc., Boston, pp. 401–420.

Haurwitz, B.: 1956, 'The geographical distribution of the solar semidiurnal pressure oscillation', *Meteorol. Pap.* **2** (5), New York University.

Haurwitz, B.: 1957, 'Atmospheric oscillations and meridional temperature gradient', *Beitr. Phys. Atmos.* **30**, 46–54.

Haurwitz, B.: 1962a, 'Die tägliche Periode der Lufttemperatur in Bodennähe und ihre geographische Verteilung', *Arch. Met. Geoph. Biokl.* **A12**, 426–434.

Haurwitz, B.: 1962b, 'Wind and pressure oscillations in the upper atmosphere', *Arch. Meteorol. Geophys. Biokl.* **13**, 144–165.

Haurwitz, B.: 1964, 'Tidal phenomena in the upper atmosphere', *W.M.O. Rept.*, No. 146, T.P. 69.

Haurwitz, B.: 1965, 'The diurnal surface pressure oscillation', *Archiv. Meteorol. Geophys. Biokl.* **A14**, 361–379.

Haurwitz, B. and Cowley, Ann D.: 1965, 'The lunar and solar air tides at six stations in North and Central America', *Mo. Wea. Rev.* **93**, 505–509.

Haurwitz, B. and Cowley, Ann D.: 1966, 'Lunar air tide in the Caribbean and its monthly variation', *Mo. Wea. Rev.* **94**, 303–306.

Haurwitz, B. and Cowley, Ann D.: 1967, 'New determinations of the lunar barometric tide', *Beitr. Phys. Atmos.* **40**, 243–261.

Haurwitz, B. and Cowley, Ann D.: 1968, 'Lunar and solar barometric tides in Australia', *Mo. Wea. Rev.* **96**, 601–605.

Haurwitz, B. and Cowley, A. D.: 1968a, 'Lunar tidal winds at four American stations', *Geophys. J. Roy. Astron. Soc.* **15**, 103–107.

Haurwitz, B. and Cowley, A. D.: 1969, 'Lunar semi-diurnal wind variations at Hongkong and Uppsala', *Quart. J. Roy. Meteorol. Soc.* **95**, 766–770.

Haurwitz, B. and Cowley, A. D.: 1970, 'The lunar barometric tide, its global distribution and annual variation', *Pure and Applied Geophys.* **75**, 1–29.

Haurwitz, B. and Möller, F.: 1955, 'The semidiurnal air-temperature variation and the solar air tide', *Archiv. Meteorol. Geophys. Biokl.* **A8**, 332–350.

Haurwitz, B. and Sepúlveda, G. M.: 1957, 'The geographical distribution and seasonal variation of the semidiurnal pressure oscillation in high latitudes', *Archiv Meteorol. Geophys. Biokl.*, **A10**, 29–42.

Hering, W. S. and Borden, T. R.: 1962, 'Diurnal variations in the summer wind field over the central United States', *J. Atmos. Sci.* **19**, 81–86.

Hines, C. O.: 1960, 'Internal gravity waves at ionospheric heights', *Canad. J. Phys.* **38**, 1441–1481.

Hines, C. O.: 1963, 'The upper atmosphere in motion', *Quart. J. Roy. Meteorolog. Soc.* **89**, 1–42.

Hines, C. O.: 1966, 'Diurnal tide in the upper atmosphere', *J. Geophys. Res.* **71**, 1453–1459.

Hodges, R. R. Jr.: 1967, 'Generation of turbulence in the upper atmosphere by internal gravity waves', *J. Geophys. Res.* **72**, 3455–3458.

Holmberg, E. R. R.: 1952, 'A suggested explanation of the present value of the velocity of rotation of the earth', *Monthly Notices Roy. Astron. Soc. Geophys. Suppl.* **6**, 325–330.

Hough, S. S.: 1897, 'On the application of harmonic analysis to the dynamical theory of tides, Part I. On Laplace's 'Oscillations of the first species', and on the dynamics of ocean currents', *Phil. Trans. Roy. Soc.* **A189**, 201–257.

Hough, S. S.: 1898, 'The application of harmonic analysis to the dynamical theory of the tides, Part II. On the general integration of Laplace's dynamical equations', *Phil. Trans. Roy. Soc. London* **A191**, 139–185.

Hunt, D. C. and Manabe, S.: 1968, 'An investigation of thermal tidal oscillations in the earth's atmosphere using a general circulation model', *Mon. Wea. Rev.* **96**, 753–766.

Hyde, W. W.: 1947, *Ancient Greek Mariners*, Oxford University Press, New York.

Hylleraas, E. A.: 1939, 'Über die Schwingungen eines stabil geschichteten, durch Meridiane begrenzten Meeres, I', *Astrophys. Norveg.* **3**, 139–164.

Hyson, P.: 1968, 'Thermistor mountings', *J. App. Meteorol.* **7**, 908–918.

Jacchia, L. G.: 1963, 'Variations of the earth's upper atmosphere as revealed by satellite drag', *Rev. Mod. Phys.* **35**, 973–991.

Jacchia, L. G. and Kopal, Z.: 1951, 'Atmospheric oscillations and the temperature profile of the upper atmosphere', *J. Meteorol.* **9**, 13–23.

James, G. and James, R. C.: 1959, *Mathematics Dictionary*, Van Nostrand.

Johnson, D. H.: 1955, 'Tidal oscillations of the lower stratosphere', *Quart. J. Roy. Meteorol. Soc.* **81**, 1–8.

Johnson, F. S.: 1953, 'High altitude diurnal temperature changes due to ozone absorption', *Bull. Am. Meteorol. Soc.* **34**, 106–110.

Kato, S.: 1956, 'Horizontal wind systems in the ionospheric E region deduced from the dynamo theory of geomagnetic Sq variation, Part II', *J. Geomagnet. Geoelec. Kyoto* **8**, 24–37.

Kato, S.: 1966, 'Diurnal atmospheric oscillation, 1, eigenvalues and Hough functions', *J. Geophys. Res.* **71**, 3201–3209.

Kelvin: see Thomson.

Kertz, W.: 1951, 'Theorie der gezeitenartigen Schwingungen als Eigenwertproblem', *Ann. Meteorol.* **4**, Suppl. 1.

Kertz, W.: 1956a, 'Die thermische Erregungsquelle der atmospharischen Gezeiten', *Nachr. Akad Wiss. Göttingen Math.-phys. Kl.* No. 6, 145–166.

Kertz, W.: 1956b, 'Components of the semidiurnal pressure oscillation', New York University, Dept. of Meteor. and Ocean., Sci. Rep. 4.

Kertz, W.: 1956c, 'The seasonal variations of the six-hourly planetary pressure and temperature waves', New York Univ., Dept. of Meteor. and Ocean, Sci. Rep. 5.

Kertz, W.: 1957, 'Atmosphärische Gezeiten', in *Handbuch der Physik* (ed. by S. Flügge), **48**, 928–981, Springer, Berlin.

Kertz, W.: 1959, 'Partialwellen in den halb- und vierteltägigen gezeitenartigen Schwingungen der Erdatmosphäre', *Arch. Meteorol. Geophys. Biokl.* **A11**, 48–63.

King, J. W. and Kohl, H.: 1965, 'Upper atmospheric winds and ionospheric drifts caused by neutral air pressure gradients', *Nature* **206**, 699–701.

King-Hele, D. G., and Walker, D. M. C.: 1961, 'Upper-atmosphere density during the years 1957 to 1961, determined from satellite orbits', *Space Res.* **2**, 918–957.

Kiser, W. L., Carpenter, T. H., and Brier, G. W.: 1963, 'The atmospheric tides at Wake Island', *Mo. Wea. Rev.* **91**, 566–572.

Kondratyev, K. Ya.: 1965, *Radiative Heat Exchange in the Atmosphere*, Pergamon Press, London.

Kuo, H. L.: 1968, 'The thermal interaction between the atmosphere and the earth and propagation of diurnal temperature waves', *J. Atmos. Sci.* **25**, 682–706.

Lamb, H.: 1910, 'On atmospheric oscillations', *Proc. Roy. Soc.* **A84**, 551–572.

Lamb, H.: 1932, *Hydrodynamics*, Cambridge University Press, Cambridge, England. 4th edition 1916; 5th edition, 1924; 6th edition, 1932.

Lamont, J.: 1868, Ann. Astron. Observ. Munich, Suppl. vol. 6.

Landau, L. D. and Lifshitz, E. M.: 1959, *Fluid Mechanics*, Pergamon Press, London.

Laplace, P. S. (later Marquis De La Place): 1799, *Mécanique céleste*, Paris (a) 2 (iv), 294–298.

Laplace, P. S. (later Marquis De La Place): 1825, *Mécanique céleste*, Paris, (b) 5 (xiii), 145–167; (c) 5 (xiii), 237–243; (d) 5 (Supp.), 20–35 (dated 1827, but published after Laplace's death in 1830); (e) 5 (xii), 95. The substance of (b) and (c) was taken from 'De l'action de la lune sur l'atmosphère', *Ann. Chim. (Phys.)*, **24** (1823), 280–294, and was reviewed, and partly translated, in 'Berechnung der von dem Monde bewirkten atmosphärischen Fluth', *Pogg. Ann. Phys. Chem.* **13** (1828), 137–149. Laplace also contributed 'Additions' on this subject to the *Connaissance des Tems (sic)* for 1826 and 1830.

Lenhard, R. W.: 1963, 'Variation of hourly winds at 35 to 65 km during one day at Eglin Air Force Base, Florida', *J. Geophys. Res.* **68**, 227–234.

Leovy, C.: 1964, 'Radiative equilibrium of the mesosphere', *J. Atmos. Sci.* **21**, 238–248.

Liller, W. and Whipple, F. L.: 1954, 'High altitude winds by meteor-train photography', Spec. Suppl. to *J. Atmos. Terr. Phys.* **1**, 112–130.

Lindzen, R. S.: 1966a, 'On the theory of the diurnal tide', *Mon. Wea. Rev.* **94**, 295–301.

Lindzen, R. S.: 1966b, 'On the relation of wave behavior to source strength and distribution in a propagating medium', *J. Atmos. Sci.* **23**, 630–632.

Lindzen, R. S.: 1967a, 'Thermally driven diurnal tide in the atmosphere', *Quart. J. Roy. Meteorol. Soc.* **93**, 18–42.

Lindzen, R. S.: 1967b, 'Planetary waves on beta-planes', *Mon. Wea. Rev.* **95**, 441–451.

Lindzen, R. S.: 1967c, 'Lunar diurnal atmospheric tide', *Nature* **215**, 1260–1261.

Lindzen, R. S.: 1967d, 'Reconsideration of diurnal velocity oscillation in the thermosphere', *J. Geophys. Res.* **72**, 1591–1598.

Lindzen, R. S.: 1967e, 'Physical processes in the mesosphere', Proceedings of the IAMAP Moscow meeting on *Dynamics of Large Scale Atmospheric Processes* (ed. by A. S. Monin).

Lindzen, R. S.: 1968a, 'The application of classical atmospheric tidal theory', *Proc. Roy. Soc.* **A303**, 299–316.

Lindzen, R. S.: 1968b, 'Vertically propagating waves in an atmosphere with Newtonian cooling inversely proportional to density', *Can. J. Phys.* (in press).

Lindzen, R. S., Batten, E. S., and Kim, J.-W.: 1968, 'Oscillations in atmospheres with tops', *Mon. Wea. Rev.* **96**, 133–140.

Lindzen, R. S. and Goody, R.: 1965, 'Radiative and photochemical processes in mesospheric dynamics. Part I: Models for radiative and photochemical processes', *J. Atmos. Sci.* **22**, 341–348.

Lindzen, R. S. and McKenzie, D. J.: 1967, 'Tidal theory with Newtonian cooling', *Pure App. Geophys.* **66**, 90–96.

Longuet-Higgins, M. S.: 1967, 'The eigenfunctions of Laplace's tidal equations over a sphere', *Phil. Trans. Roy. Soc.* **A269**, 511–607.

Love, A. E. H.: 1913, 'Notes on the dynamical theory of the tides', *Proc. London Math. Soc.* **12**, 309–314.

Maeda, H.: 1955, 'Horizontal wind systems in the ionospheric E-region deduced from the dynamo theory of the geomagnetic S_q variation, Part I', *J. Geomagnet. Geoelec. Kyoto* **7**, 121–132.

Manabe, S., and Möller, F.: 1961, 'On the radiative equilibrium and heat balance of the atmosphere', *Mon. Wea. Rev.* **89**, 503–532.

Manring, E., Bedinger, J., Knaflich, H., and Layzer, D.: 1964, 'An experimentally determined model for the periodic character of winds from 85–135 km', NASA Contractor Rept., NASA CR-36.

Margules, M.: 1890, 'Über die Schwingungen periodisch erwarmter Luft', *Sitzber. Akad. Wiss. Wien, Abt.* IIa, **99**, 204–227.

Margules, M.: 1892, 'Luftbewegungen in einer rotierenden Sphäroidschale', *Sitzber. Akad. Wiss. Wien, Abt.* IIa, **101**, 597–626.

Margules, M.: 1893, *Sitzber. Akad. Wiss. Wien, Abt.* IIa, **102**, 11–56; 1369–1421.

Martyn, D. F.: 1955, 'Interpretation of observed F_2 winds as ionization drifts associated with the magnetic variations', *The Physics of the Ionosphere*, Report of the Physical Society, London, 163–165.

Martyn, D. F. and Pulley, O. O.: 1947, 'Atmospheric tides in the ionosphere: Part II, Lunar tidal variations in the F-region near the magnetic equator', *Proc. Roy. Soc.* **A190**, 273–288.

Matsushita, S. and Campbell, W. H.: 1967, *Physics of Geomagnetic Phenomena*, Vol. I, Academic Press, New York and London (2 vols.).

Midgely, J. E. and Liemohn, H. B.: 1966, 'Gravity waves in a realistic atmosphere', *J. Geophys. Res.* **71**, 3729–3748.

Miers, B. T.: 1965, 'Wind oscillations between 30 and 60 km over White Sands Missile Range, New Mexico', *J. Atmos. Sci.* **22**, 382–387.

Minzner, R. A., Champion, K. S. W., and Pond, H. L.: 1959, *The ARDC Model Atmosphere, 1959*. (Air Force Surveys in Geophysics, No. 115.)

Möller, F.: 1940, 'Über den Tagesgang des Windes', *Meteor. Z.* **57**, 324–331.

Morano, F.: 1899, 'Marea atmosferica', *R. C. Accad., Lincei* **8**, 521–528.

Mügge, R. and Möller, F.: 1932, 'Zur Berechnung von Strahlungsströmen und Temperaturänderungen in Atmosphären von beliebigen Aufbau', *Z. Geophysik* **8**, 53–64.

Nawrocki, P. J. and Papa, R.: 1963, *Atmospheric Processes*, Prentice-Hall, New Jersey.

Neamtan, S. M.: 1946, 'The motion of harmonic waves in the atmosphere', *J. Meteorol.* **3**, 53–56.

Neumayer, G.: 1867, 'On the lunar atmospheric tide at Melbourne', *Proc. Roy. Soc.* **15**, 489–501.

Newton, I.: 1687, *Philosophiae Naturalis Principia Mathematica*, (a) Bk. 1, Prop. 66, Cor. 19, 20; Bk. 3, Prop. 24, 36, 37, (b) Bk. 2, Prop. 48–50.

Newton, I.: 1727, *De Mundi Systemate*, London, Sections 38–47, 49–54.

Nunn, D.: 1967, *A theoretical study of tides in the upper atmosphere*, M.Sc. thesis, McGill University, Montreal.

Palumbo, A.: 1960, 'La marea atmosferica lunare a Catania', *Atti. Assoc. Geofis. Italia.*

Palumbo, A.: 1962, 'La marea atmosferica lunare a Napoli', *Atti. Assoc. Geofis. Italia.*

Pekeris, C. L.: 1937, 'Atmospheric oscillations', *Proc. Roy. Soc.* **A158**, 650–671.

Pekeris, C. L.: 1939, 'The propagation of a pulse in the atmosphere', *Proc. Roy. Soc.* **A171**, 434–449.

Pekeris, C. L.: 1951, 'Effect of the quadratic terms in the differential equations of atmospheric oscillations', Natl. Adv. Comm. Aeronaut. Tech. Notes, *2314*.

Pekeris, C. L. and Alterman, Z.: 1959, 'A method of solving nonlinear equations of atmospheric tides with applications to an atmosphere of constant temperature', in *The Atmosphere and the Sea in Motion*, Rockefeller Institute Press, New York.

Phillips, N. A.: 1966, 'The equations of motion for a shallow rotating atmosphere and the 'traditional approximation'', *J. Atmos. Sci.* **23**, 626–628.

Phillips, N. A.: 1968, 'Reply to 'comments on Phillips' simplification of the equations of motion', by G. Veronis', *J. Atmos. Sci.* **25**, 1155–1157.

Pitteway, M. L. V. and Hines, C. O.: 1963, 'The viscous damping of atmospheric gravity waves', *Can. J. Phys.* **41**, 1935.

Platzman, G. W.: 1967, 'A retrospective view of Richardson's book on weather prediction', *Bull. Amer. Meteor. Soc.* **48**, 514–550; see p. 539.

Pramanik, S. K.: 1926, 'The six-hourly variations of atmospheric pressure and temperature', *Mem. Roy. Meteorol. Soc. London* 1, 35–57.

Pressman, J.: 1955, 'Diurnal temperature variations in the middle atmosphere', *Bull. Amer. Meteorol. Soc.* 36, 220–223.

Price, A. T.: 1969, 'Daily variations of the geomagnetic field', *Space Sci. Rev.* 9, 151–197.

Rayleigh, 3rd Baron (Strutt, J. W.): 1890, 'On the vibrations of an atmosphere', *Phil. Mag.* (5) 29, 173–180; Scientific Papers 3, 333–340, Dover Publications, New York, 1964.

Reed, R. J.: 1967, 'Semidiurnal tidal motions between 30 and 60 km', *J. Atmos. Sci.* 24, 315–317.

Reed, R. J., McKenzie, D. J., and Vyverberg, Joan C.: 1966a, 'Further evidence of enhanced diurnal tidal motions near the stratopause', *J. Atmos. Sci.* 23, 247–251.

Reed, R. J., McKenzie, D. J., and Vyverberg, Joan C.: 1966b 'Diurnal tidal motions between 30 and 60 km in summer', *J. Atmos. Sci.* 23, 416–423.

Reed, R. J., Oard, M. J., and Sieminski, Marya: 1969, 'A comparison of observed and theoretical diurnal tidal motions between 30 and 60 km', *Mon. Wea. Rev.* 97, 456–459.

Revah, I., Spizzichino, A., and Massebeuf, Mme.: 1967, 'Marée semi-diurnelle et vents dominants zonaux mesurés à Garchy (France) de Novembre 1965 à Avril 1966', Note Technique GRI/NTP/27 de Centre National d'Études des Télécommunications, Issy-les-Moulineaux, France.

Richtmyer, R.: 1957, *Difference methods for initial value problems*, Interscience, New York.

Robb, R. A.; see Chapman, S.: 1936, 'The lunar atmospheric tide at Glasgow', *Proc. Roy. Soc. Edin.* 56, 1–5.

Robb, R. A. and Tannahill, T. R.: 1935, 'The lunar atmospheric pressure inequalities at Glasgow', *Proc. Roy. Soc. Edin.* 55, 91–96.

Roberts, P. H.: 1967, *An Introduction to Magnetohydrodynamics*, American Elsevier, New York.

Rodgers, C. D. and Walshaw, C. D.: 1966, 'The computation of infrared cooling rate in planetary atmospheres', *Quart. J. Roy. Meteorol. Soc.* 92, 67–92.

Rooney, W. J.: 1938, 'Lunar diurnal variation in earth currents at Huancayo and Tucson', *Terr. Magn. Atmos. Elec.* 43, 107–118.

Rosenberg, N. W. and Edwards, H. D.: 1964, 'Observations of ionospheric wind patterns through the night', *J. Geophys. Res.* 69, 2819–2826.

Rosenthal, S. L. and Baum, W. A.: 1956, 'Diurnal Variation of Surface Pressure over the North Atlantic Ocean', *Mo. Weather Rev.* 84, 379–387.

Rougerie, P.: 1957, 'La marée barométrique à Paris', *Ann. de Geophys.* 13, 203–210.

Sabine, E.: 1847, 'On the lunar atmospheric tide at St. Helena', *Phil. Trans. Roy. Soc. London* 137, 45–50.

Sawada, R.: 1954, 'The atmospheric lunar tides', New York Univ. Meteorol. Pap. 2 (3).

Sawada, R.: 1956, 'The atmospheric lunar tides and the temperature profile in the upper atmosphere', *Geophys. Mag.* 27, 213–236.

Sawada, R.: 1965, 'The possible effect of oceans on the atmospheric lunar tide', *J. Atmos. Sci.* 22, 636–643.

Sawada, R.: 1966, 'The effect of zonal winds on the atmospheric lunar tide', *Arch. Meteorol. Geophys. Biokl.* A15, 129–167.

Schmidt, A.: 1890, 'Über die doppelte tägliche Oscillation des Barometers', *Meteor. Z.* 7, 182–185.

Schmidt, A.: 1919, 'Zur dritteltägigen Luftdruckschwankung' (from a letter to von Hann), *Meteor. Z.* 36, 29.

Schmidt, A.: 1921, 'Die Veranschaulichung der Resonanztheorie' (in a review of Hann's 1919 paper) *Meteor. Z.* 38, 303–304.

Schmidt, A.: 1935, *Tafeln der normierten Kugelfunktionen, sowie Formeln zur Entwicklung*, Engelhard-Reyer, Gotha.

Schou, G.: 1939, 'Mittel und Extreme des Luftdruckes in Norwegen', *Geofys. Publ.* 14, No. 2.

Schuster, A.: 1889, 'The diurnal variation of terrestrial magnetism', *Phil. Trans. Roy. Soc., London* A180, 467–518.

Sellick, N. P.: 1948, 'Note on the diurnal and semidiurnal pressure variation in Rhodesia', *Quar. J Roy. Met. Soc.* 74, 78–81.

Sen, H. K. and White, M. L.: 1955, 'Thermal and gravitational excitation of atmospheric oscillations', *J. Geophys. Res.* 60, 483–495.

Shaw, W. N.: 1936, *Manual of Meteorology*, Vol. 2, *Comparative Meteorology*, Cambridge University Press.

Siebert, M.: 1954, 'Zur theorie der thermischen Erregung gezeitenartiger Schwingungen der Erd-atmosphäre', *Naturwissenschaften* **41**, 446.

Siebert, M.: 1956a, 'Analyse des Jahresganges der 1/n-tägigen Variationen des Luftdruckes und der Temperatur', *Nachr. Akad. Wiss. Göttingen Math-phys. Kl.*, No. 6, 127–144.

Siebert, M.: 1956b, 'Über die gezeitenartigen Schwingungen der Erdatmosphäre', *Ber. Deut. Wetterd.* **4**, 65–71; 87–88.

Siebert, M.: 1957, 'Tidal oscillations in an atmosphere with meridional temperature gradient', Sci. Rept. No. 3, Project 429, N.Y. University, Dept. of Meteorol. Oceanogr.

Siebert, M.: 1961, 'Atmospheric tides', in *Advances in Geophysics*, Vol. 7, Academic Press, New York, pp. 105–182.

Simpson, G. C.: 1918, 'The twelve-hourly barometer oscillation', *Quart. J. Roy. Meteorol. Soc.* **44**, 1–18.

Solberg, H.: 1936, 'Über die freien Schwingungen einer homogen Flüssigkeitenschicht auf der rotierenden Erde I.', *Astrophys. Norweg.* **1**, 237–340.

Spar, J.: 1952, 'Characteristics of the semidiurnal pressure waves in the United States', *Bull. Amer. Meteorol. Soc.* **33**, 438–441.

Stolov, H. L.: 1954, 'Tidal wind fields in the atmosphere', *J. Meteor.* **12**, 117–140.

Sugiura, M. and Fanselau, G.: 1966, 'Lunar phase numbers v and v' for years 1850–2050', NASA Rept., X-612-66-401, Goddard Space Flight Center, Greenbelt, Maryland, USA.

Taylor, G. I.: 1917, 'Phenomena connected with turbulence in the lower atmosphere', *Proc. Roy. Soc* **A94**, 137–155.

Taylor, G. I.: 1929, 1930, 'Waves and tides in the atmosphere', *Proc. Roy. Soc.* **A126**, 169–183, 728.

Taylor, G. I.: 1932, 'The resonance theory of semidiurnal atmospheric oscillations', *Mem. Roy. Meteorol. Soc.* **4**, 41–52.

Taylor, G. I.: 1936, 'The oscillations of the atmosphere', *Proc. Roy. Soc.* **A156**, 318–326.

Thomson, W. (later Lord Kelvin): 1882, 'On the thermodynamic acceleration of the earth's rotation', *Proc. Roy. Soc. Edinb.* **11**, 396–405.

Tschu, K. K.: 1949, 'On the practical determination of lunar and lunisolar daily variations in certain geophysical data', *Australian J. Sci. Res.* **A2**, 1–24.

Van der Stok, J. P.: 1885, 'On the lunar atmospheric tide', *Obsns. Magn. Meteor. Obs. Batavia* **6**, (3)–(8), Appendix 2.

Wallace, J. M. and F. R. Hartranft: 1969, 'Diurnal wind variations; surface to 30 km', *Mon. Wea. Rev.* **96**, 446–455.

Weekes, K. and Wilkes, M. V.: 1947, 'Atmospheric oscillations and the resonance theory', *Proc. Roy. Soc.* **A192**, 80–99.

Wegener, A.: 1915, 'Zur Frage der atmosphärischen Mondgezeiten', *Meteor. Z.* **32**, 253–258.

Whipple, F. J. W.: 1918, 'A note on the propagation of the semi-diurnal pressure wave', *Quart. J. Roy. Meteorol. Soc.* **44**, 20–23.

Whipple, F. J. W.: 1930, 'The great Siberian meteor and the waves, seismic and aerial, which it produced', *Quart. J. Roy. Meteor. Soc.* **56**, 287–303.

Whittaker, E. T. and Watson, G. N.: 1927, *A Course of Modern Analysis* (4th edition), Cambridge University Press, London.

Wilkes, M. V.: 1949, *Oscillations of the Earth's Atmosphere*, Cambridge University Press.

Wilkes, M. V.: 1951, 'The thermal excitation of atmospheric oscillations', *Proc. Roy. Soc.* **A207**, 358–344.

Wilkes, M. V.: 1952, 'Worldwide oscillations of the earth's atmosphere', *Quart. J. Roy. Meteor. Soc.* **78**, 321–336.

Wilkes, M. V.: 1962, 'Oscillations of the earth's atmosphere with allowance for variation of temper-ature with latitude', *Proc. Roy. Soc.* **A271**, 44–56.

Wright, T. (ed.): 1863, A. Neckam: *De naturis rerum libri duo*, liber 2, cap. 98, Rolls Series, London.

Wulf, O. R. and Nicholson, S. B.: 1947, 'Terrestrial influences in the lunar and solar tidal motions of the air', *Terr. Magn. Atmos. Elect.* **52**, 175–182.

Wurtele, M.: 1953, The initial-value lee-wave problem for the isothermal atmosphere, U.C.L.A. Sci. Rept., Dept. of Meteorology, No. 3.

Yanowitch, M.: 1966, 'A remark on the hydrostatic approximation', *Pure Appl. Geophys.* **64**, 169–172.

Yanowitch, M.: 1967, 'Effect of viscosity on gravity waves and the upper boundary condition', *J. Fluid Mech.* **29**, 209–231.

INDEX OF NAMES

(*f* following a page number indicates reference also to the next page)

INDEX OF SUBJECTS

INDEX OF PLACES

Made in the USA
Monee, IL
14 September 2021